The Fumes of Mars

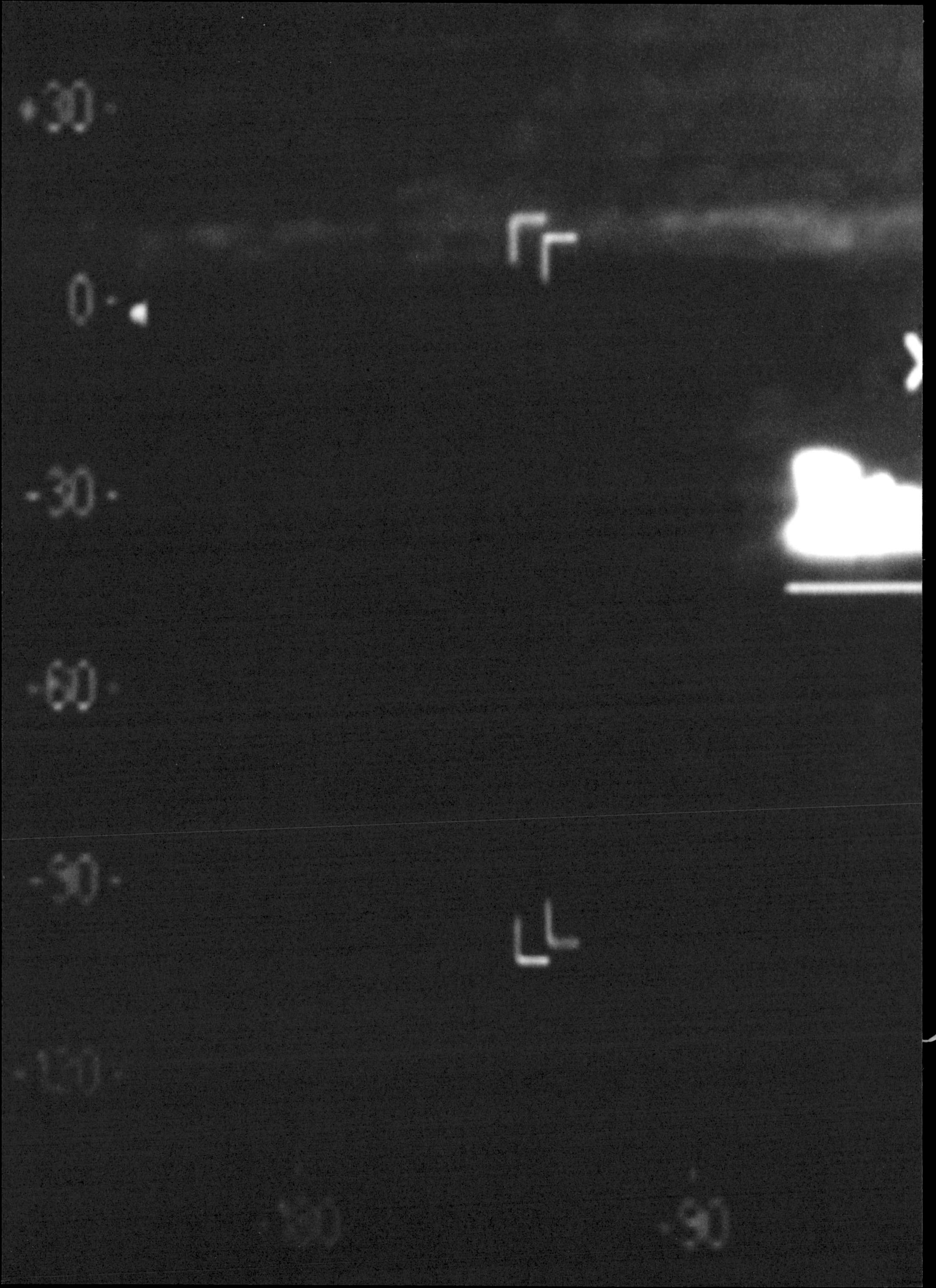
+30
0
-30
-60
-90
-120

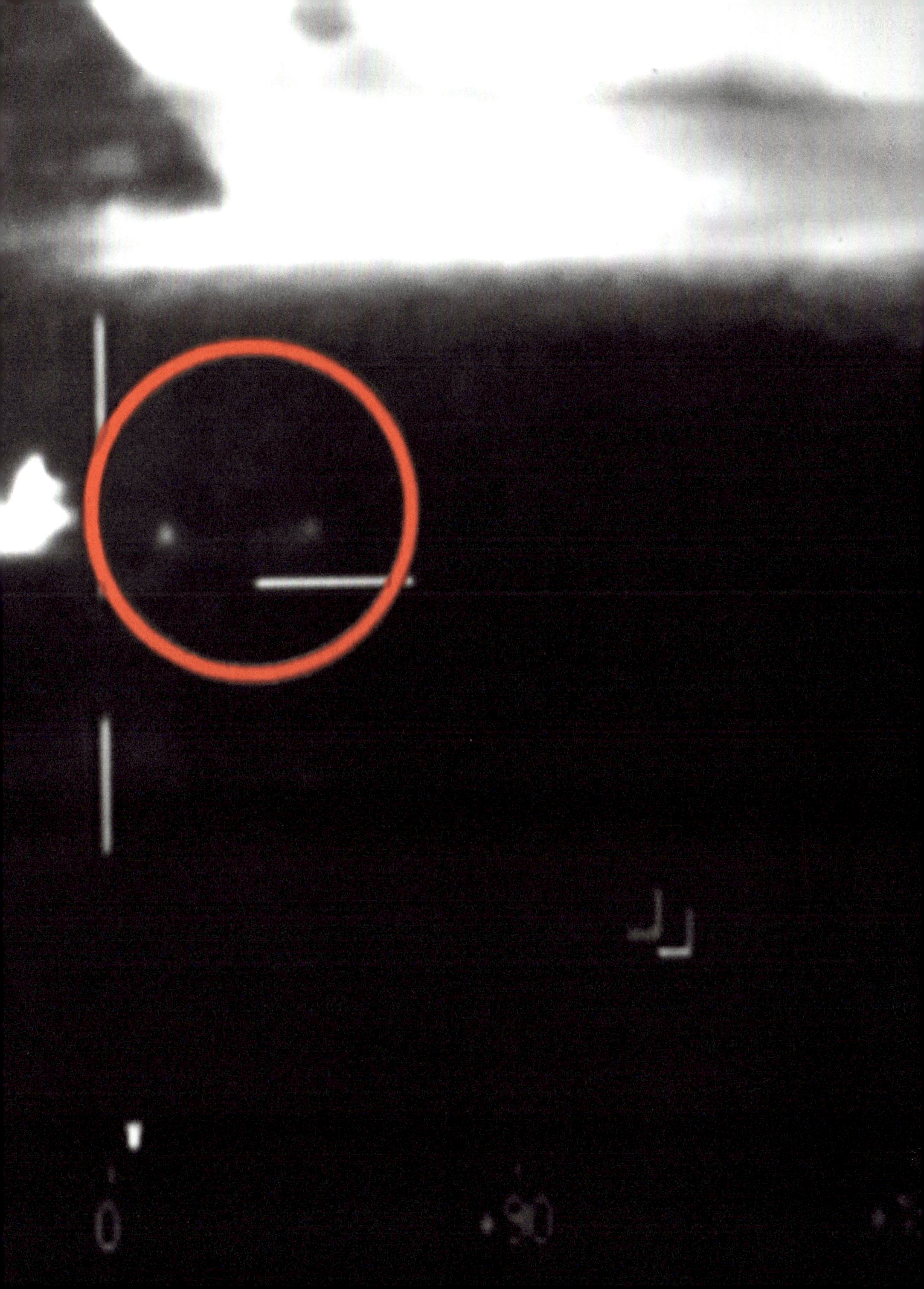

For those that perished
For those who continue to fight

For Aimi
For my Mother

For Clio

The Fumes of Mars

Katerina Angelopoulou

GOST

α'

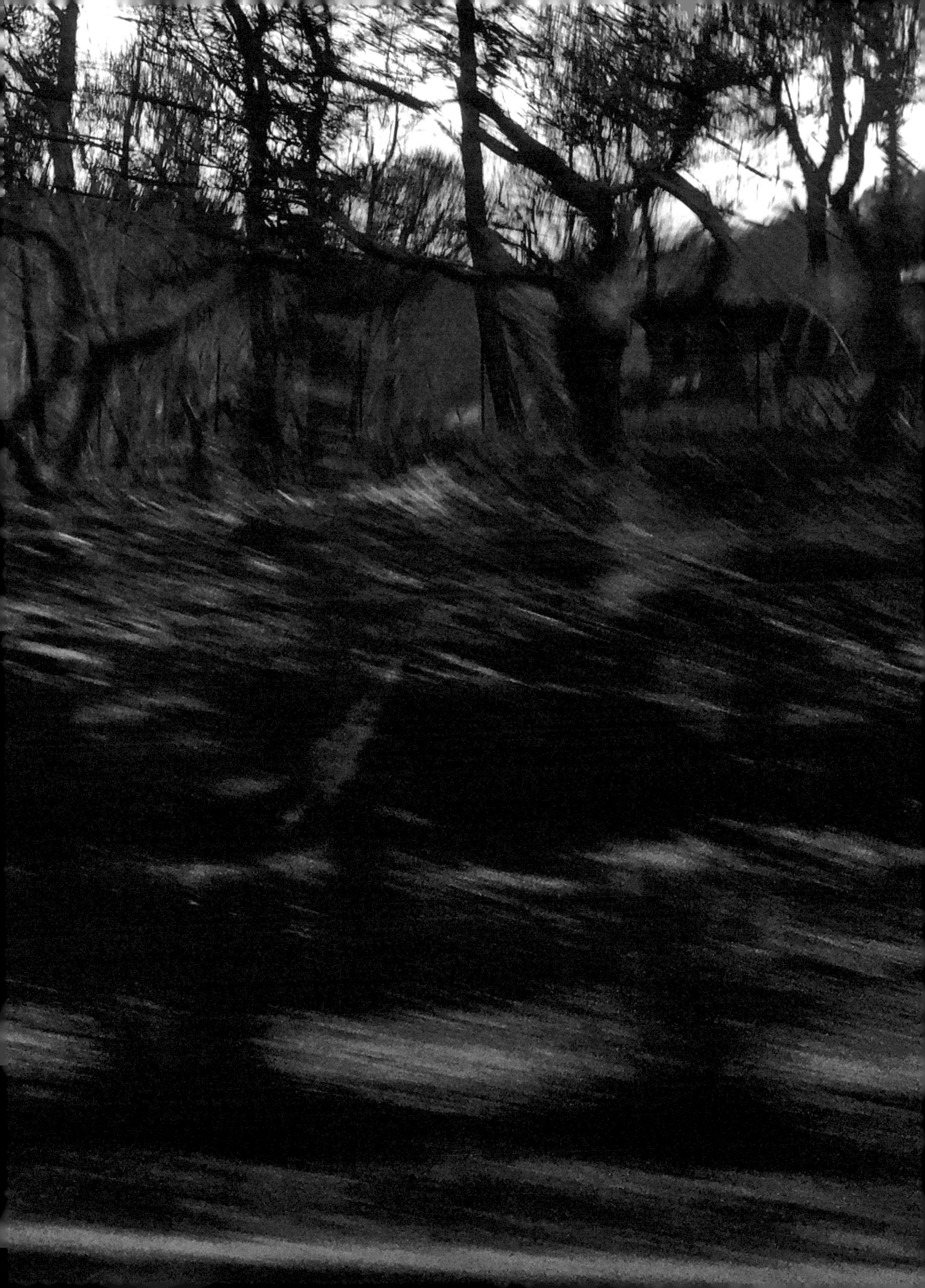

23 July 2018[1]

16:41	First report of the fire
16:50	First fire department helicopter is directed to the area with ETA 17:10
17:10	A single aircraft operates on the fire till 18.10
17:13	The fire front splits into two directions
18:06	The fire enters the village of Neos Voutzas
18:13	The fire reaches the Lyreio Foundation
18:20–18:48	No helicopter operates in inhabited areas
18:30	The fire crosses Marathonos Av.
18:30	The fire enters the village of Mati
18:45	The fire reaches the shoreline and sea

104*	Deaths (including 11 children):
71	People found dead in the area
2	Died at the nearest port
1	Baby died in the ambulance
20	Died of burns in the hospital
1	Died at home after hospitalisation
9	Drowned in the sea
172	Non-Fatal Injuries (recorded):
150	Severely burned:
57	of which with excessive burns
4,000	People directly affected
12,600 km2	Land burned
15,000	Trees burned
3,500	Damaged structures
300	Burned Cars

*On 7 April 2025 the Association of Relatives of Victims and Burn Survivors revealed there were another 14–18 confirmed unidentified victims. The bodies were never matched to any DNA, and they were never reported missing by anyone. It is assumed that they were immigrant workers.

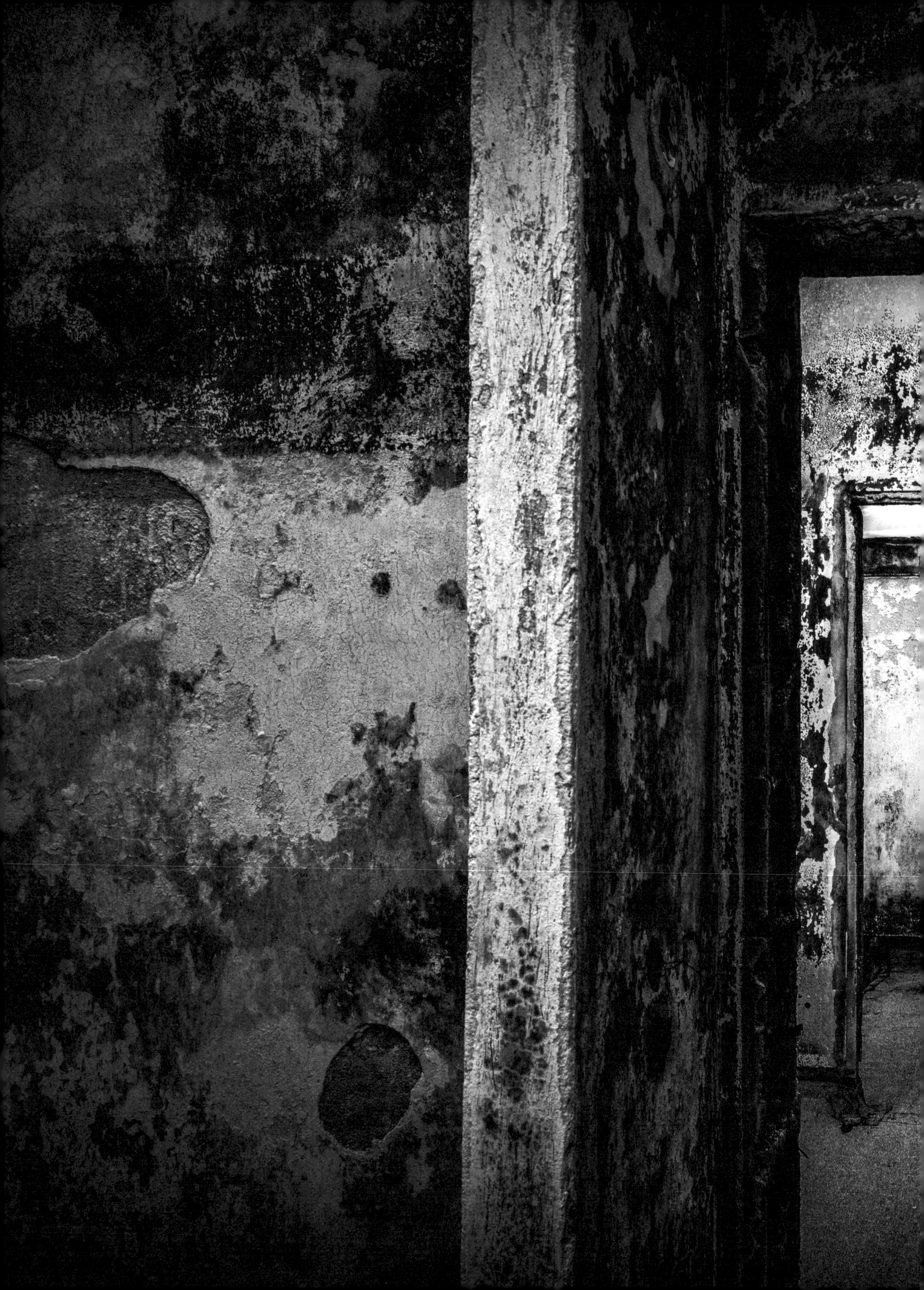

‘At 18:20 we called the police. We gave them the address and told them my wife was trapped inside, and the house was on fire...She was alive for fifty minutes...18:49 was the last phone call...Her last words to her cousin: “I know I’m going to die.”’ [2]

George Kairis

'The tragedy is the fault of the residents...' [3]
Panos Kammenos, Minister of Defense

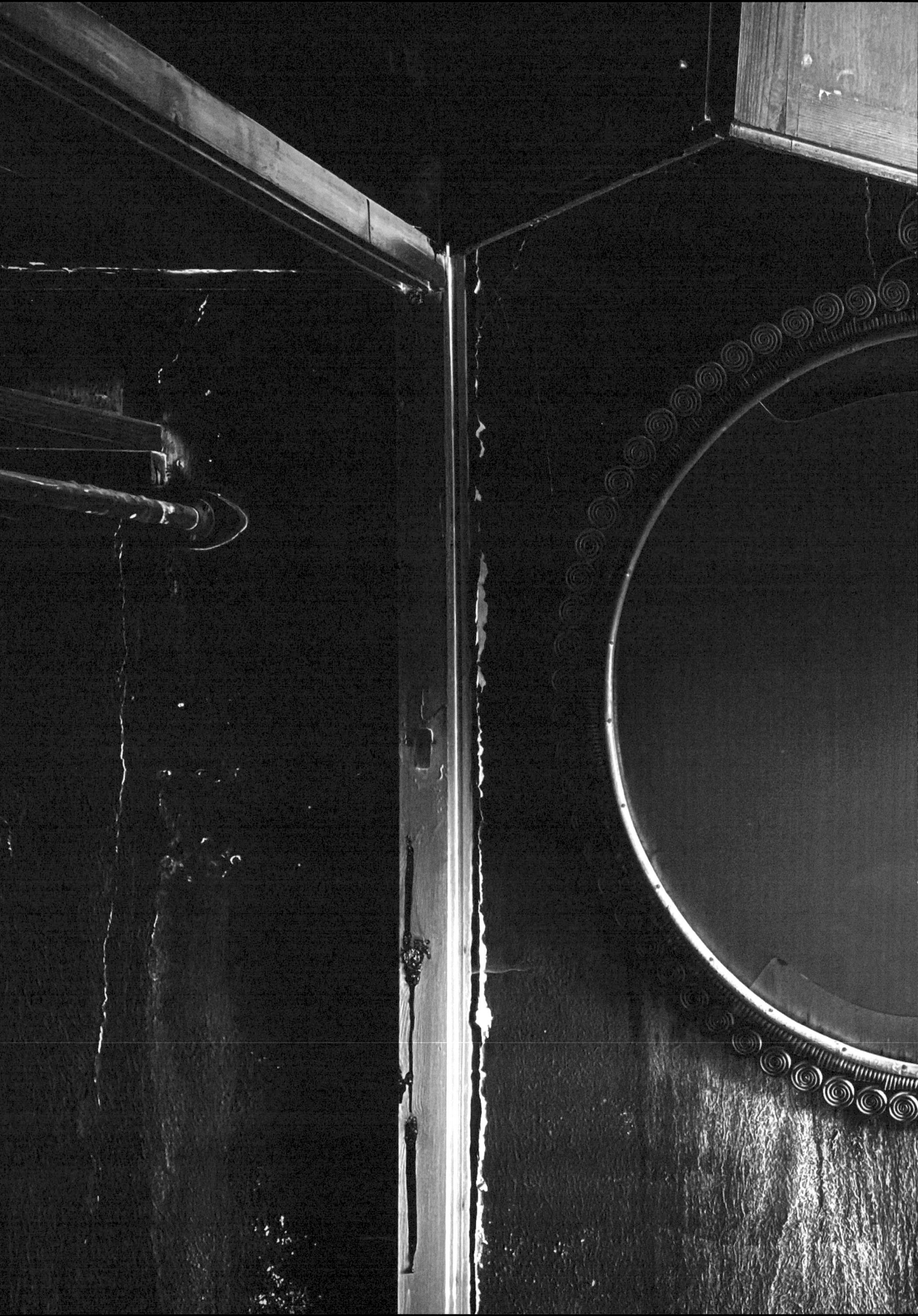

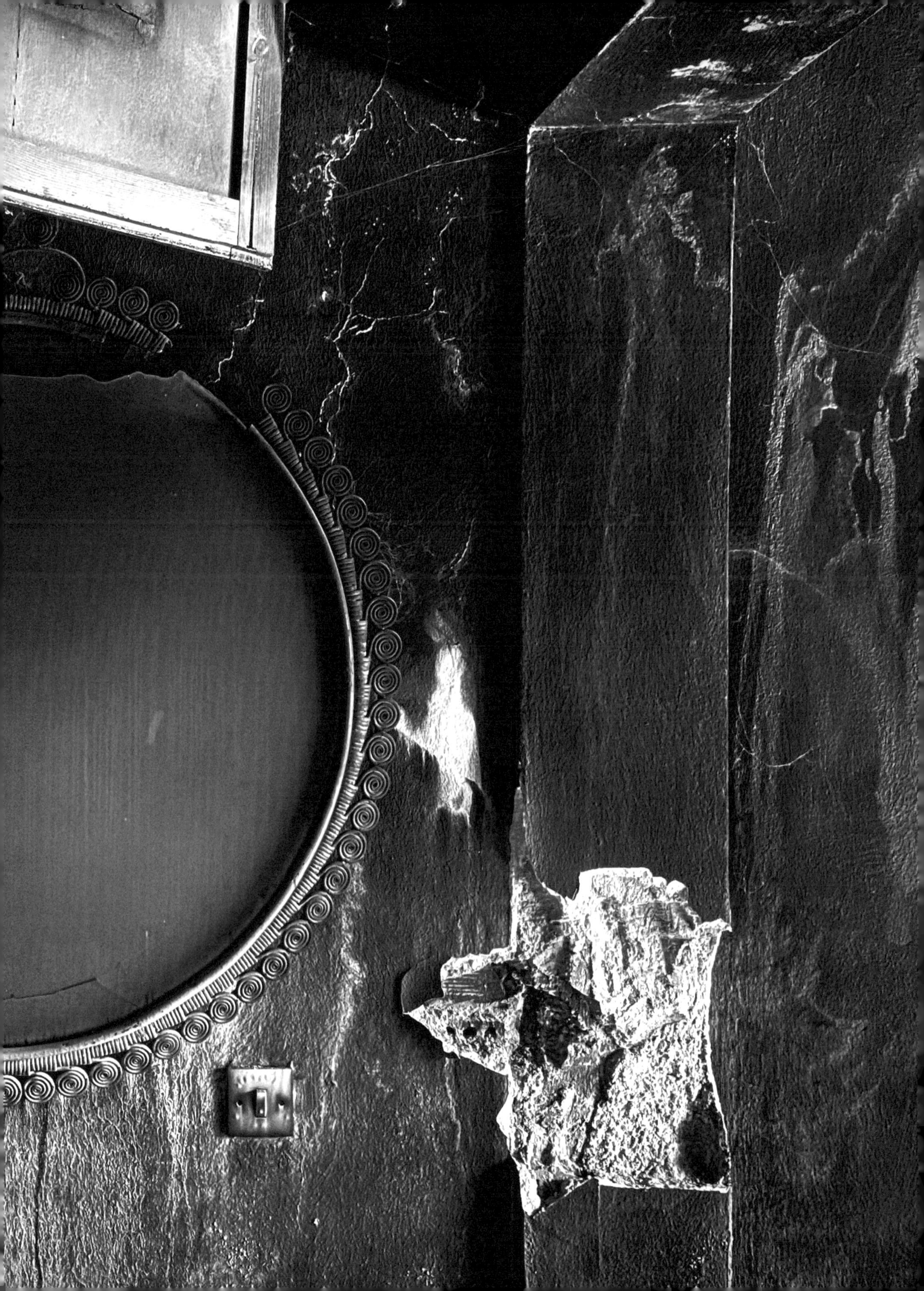

'Myself and my husband fell into the sea to save ourselves. I saw a girl—around 13-years-old—jumping off the cliff. Then I heard a scream. We went to her, but she wasn't breathing. Her little dress had caught fire and her legs were half burned. She jumped to save herself, but she didn't make it. They said she was trying to escape with her father and brother, but the fire caught up with them, and she started running.' [4]

Maria Pitsiou

I take Clio out of the car—she is just 3-years-old—and run as fast as I can. We can't see anything—the trees are on fire, pine cones exploding, embers raining down on us. The fire wasn't close a minute ago. It's faster than us. I feel I can't run anymore without tripping over. I ask Clio: 'Can you run?'

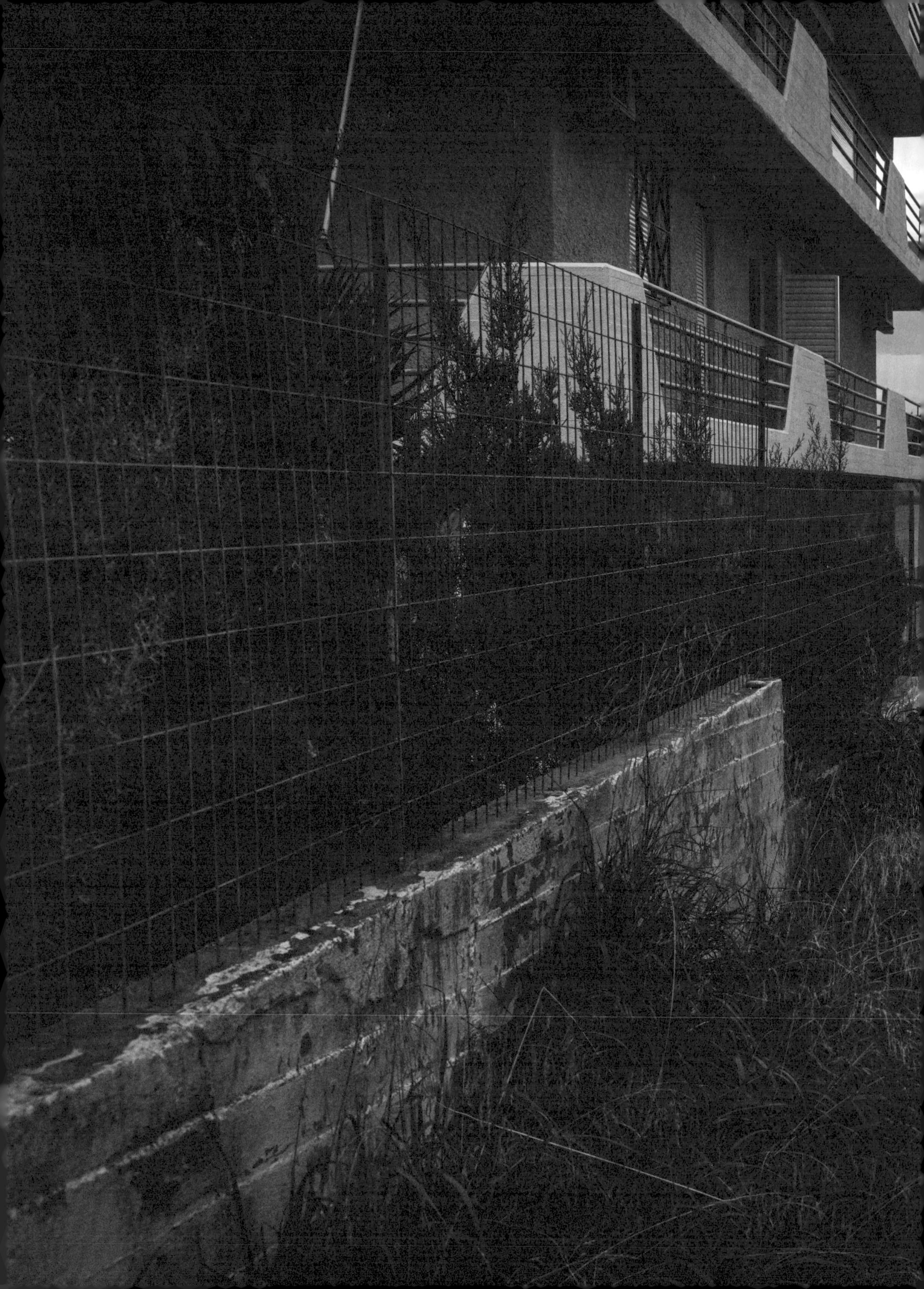

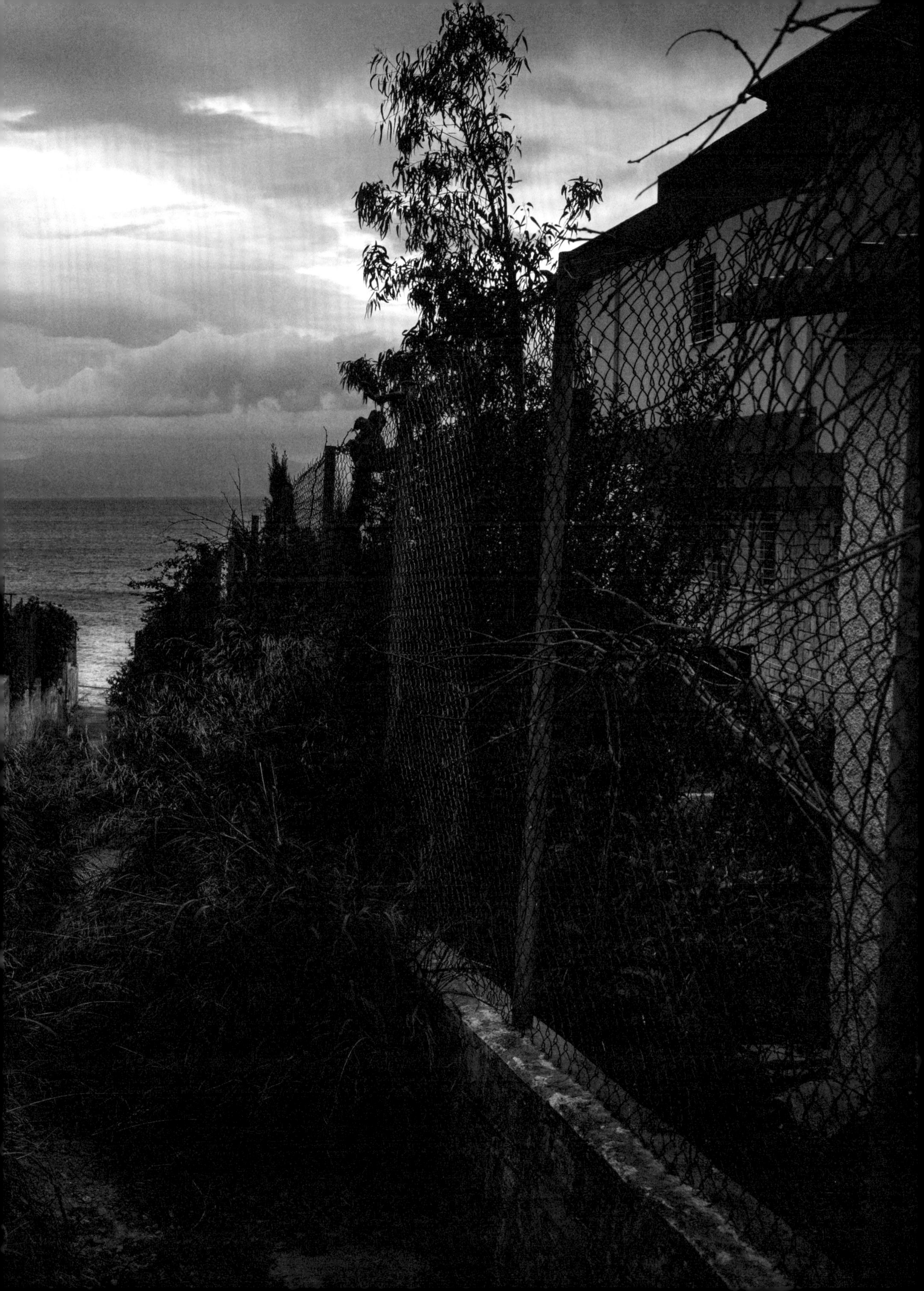

'An older woman, 10 meters (before) from the steps (that lead down to the beach), having no strength left, had surrendered to the flames and was burning alive.' [5]
Iakovos

**'After I managed to reach you, around 18:45 pm,
I called the Coast Guard in Rafina. They knew nothing.'** [6]

Margarita

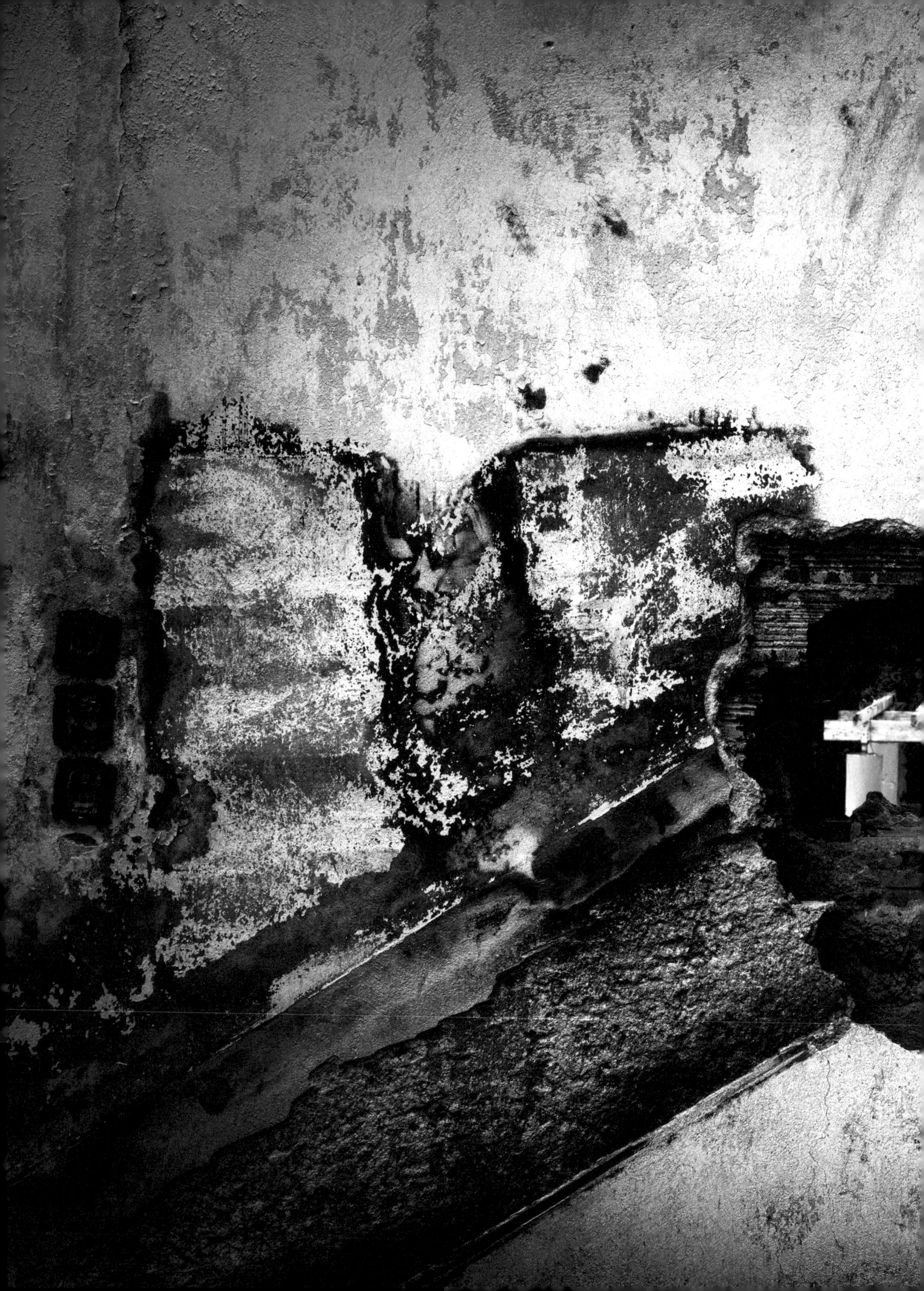

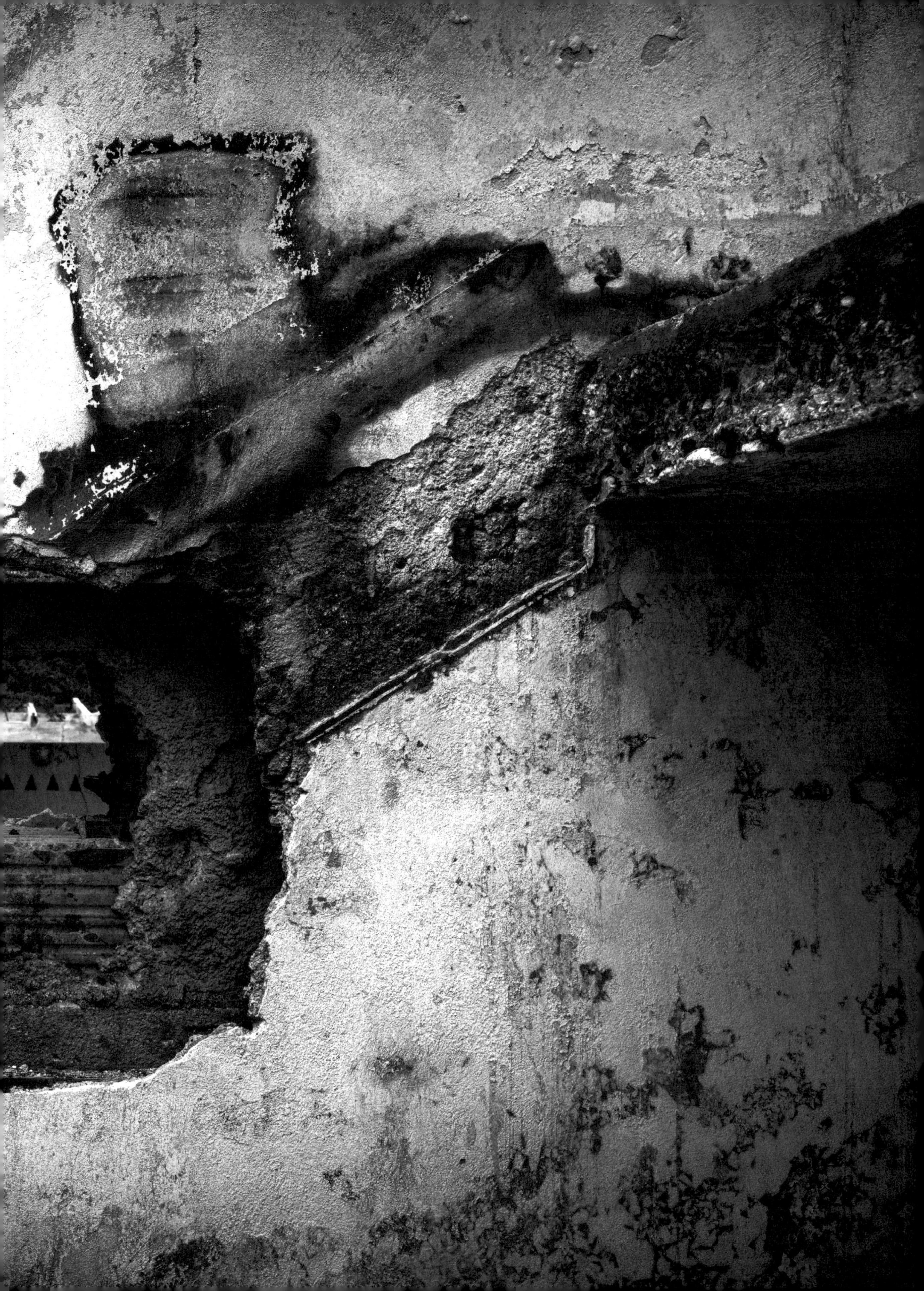

'The responsible parties, were preoccupied—obviously—with other matters [...]. And they additionally exhibited such indifference—as they did not want to be implicated actively with the operational aspect of the management of the fire in order to avoid being "blamed" for a potential failure that could compromise their promotions within the department—that they did not even order the evacuation of civilians. And they barely got involved in the rescue mission.' [1]

From the Report of Dimitris Liotsios, Fire Department Forensic Expert

'Running down this street, you had three choices: right towards Rafina, straight towards the little port of Mati, and left towards Nea Makri. One was the safe choice—left. [...] This was a life and death decision, the moment you were chased by a thermal wave of 700°C–800°C.' [2]
Christos

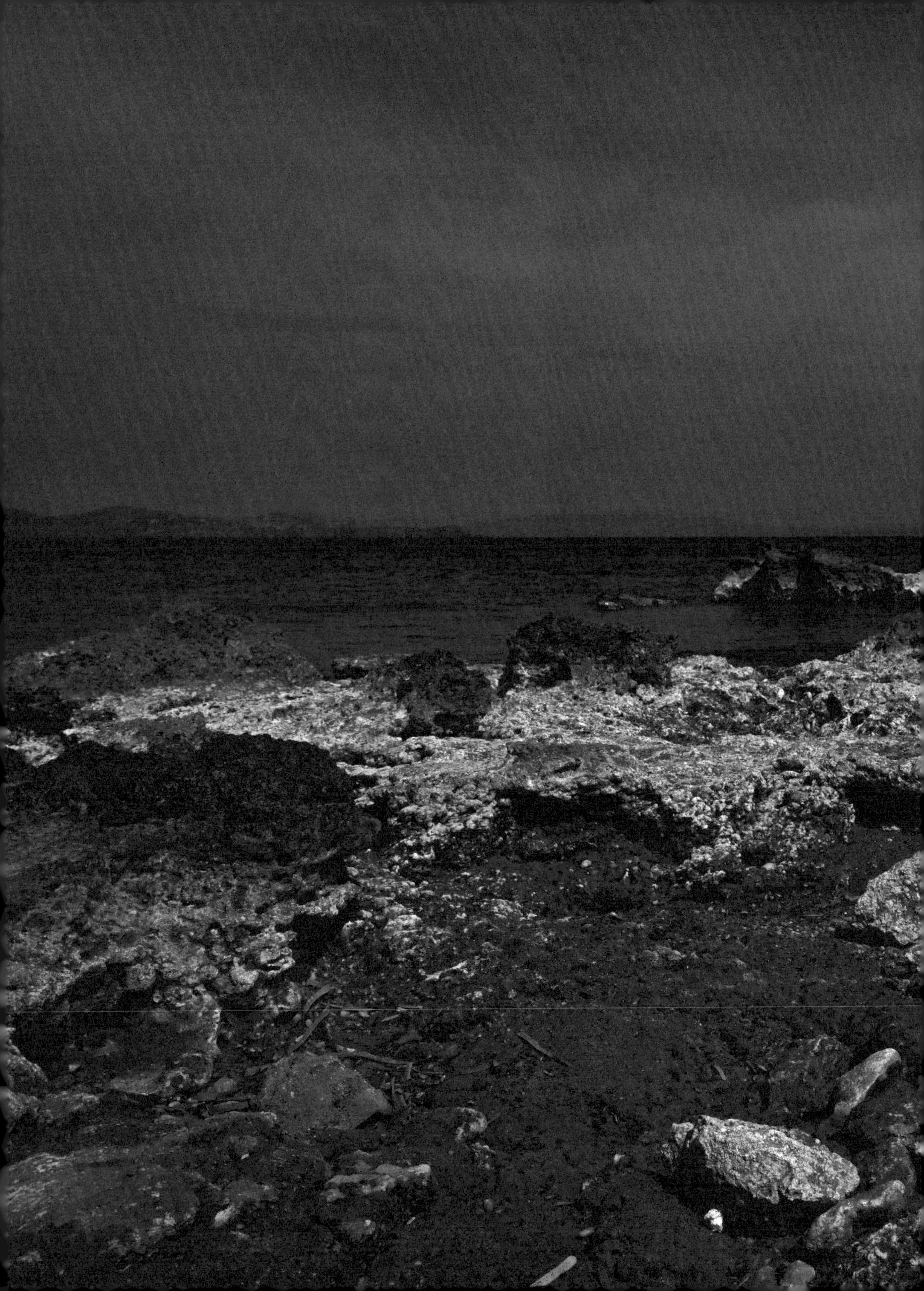

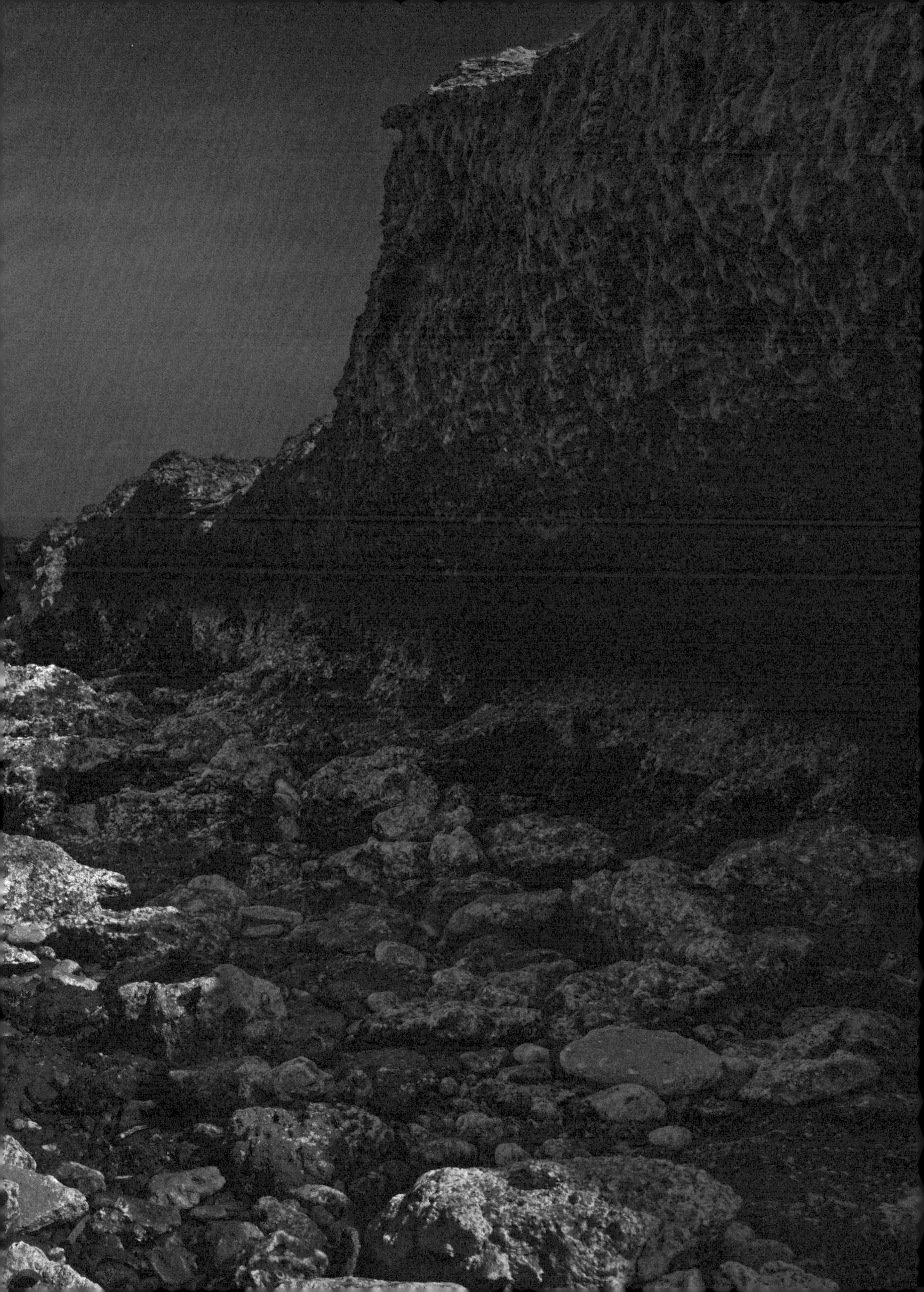

'Officials of the fire department and the Municipality, while aware of the danger of the situation, do not proceed to evacuate the area at the critical moment, i.e. between 17.30 and 17.40.' And 'At the same time, no one gave orders to the firefighting boats (four in number) but also to rescue boats of EMAK to go to the area and assist in the rescue of those who had found refuge at sea.' [7]

From the Report of Athanasios Marneris, State Investigator

'The Coast Guard said there was nothing they could do. They said they could not approach the location with their boats in order to help people because of the heavy winds and the rocky morphology of the shore.' [6]

Margarita

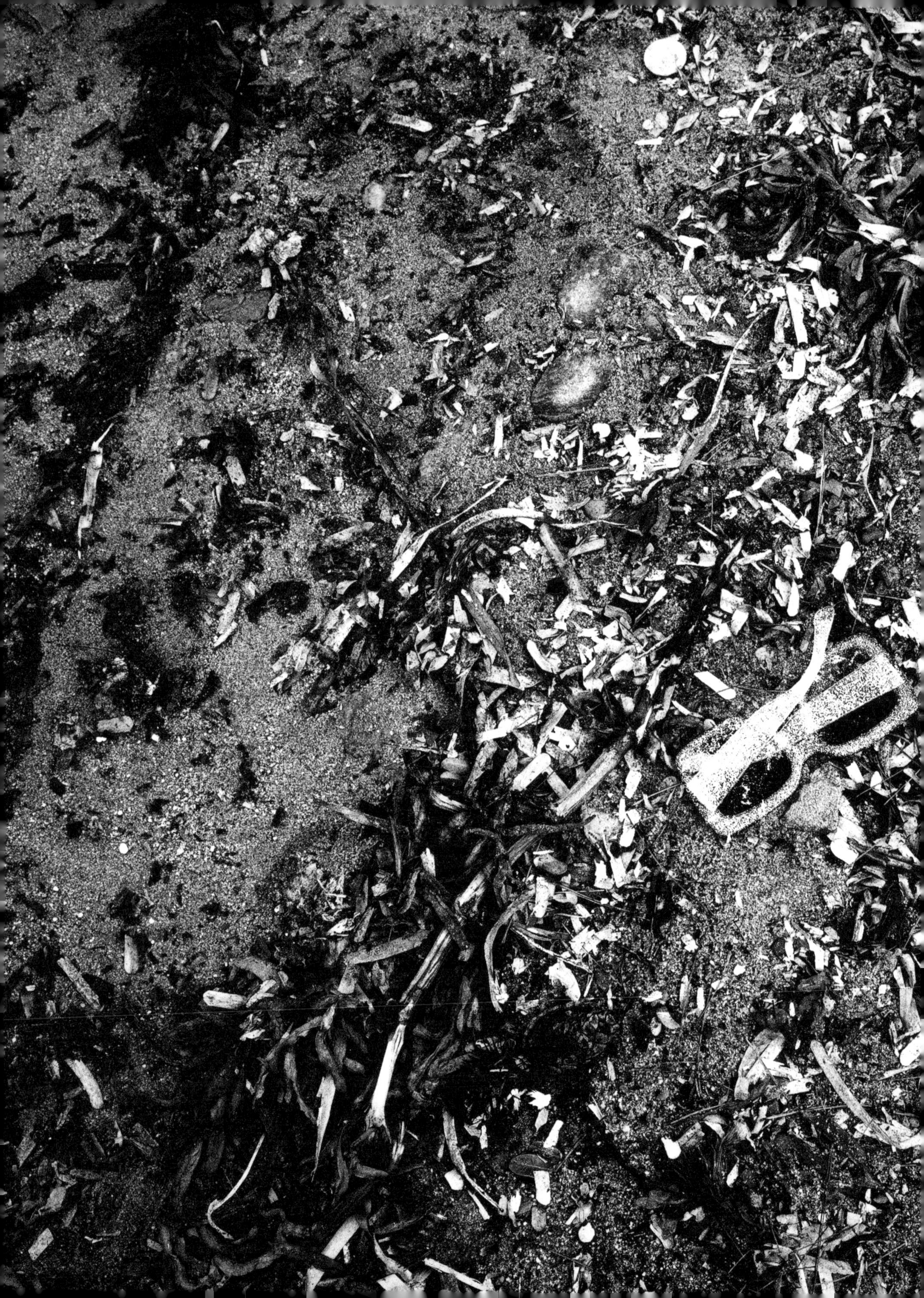

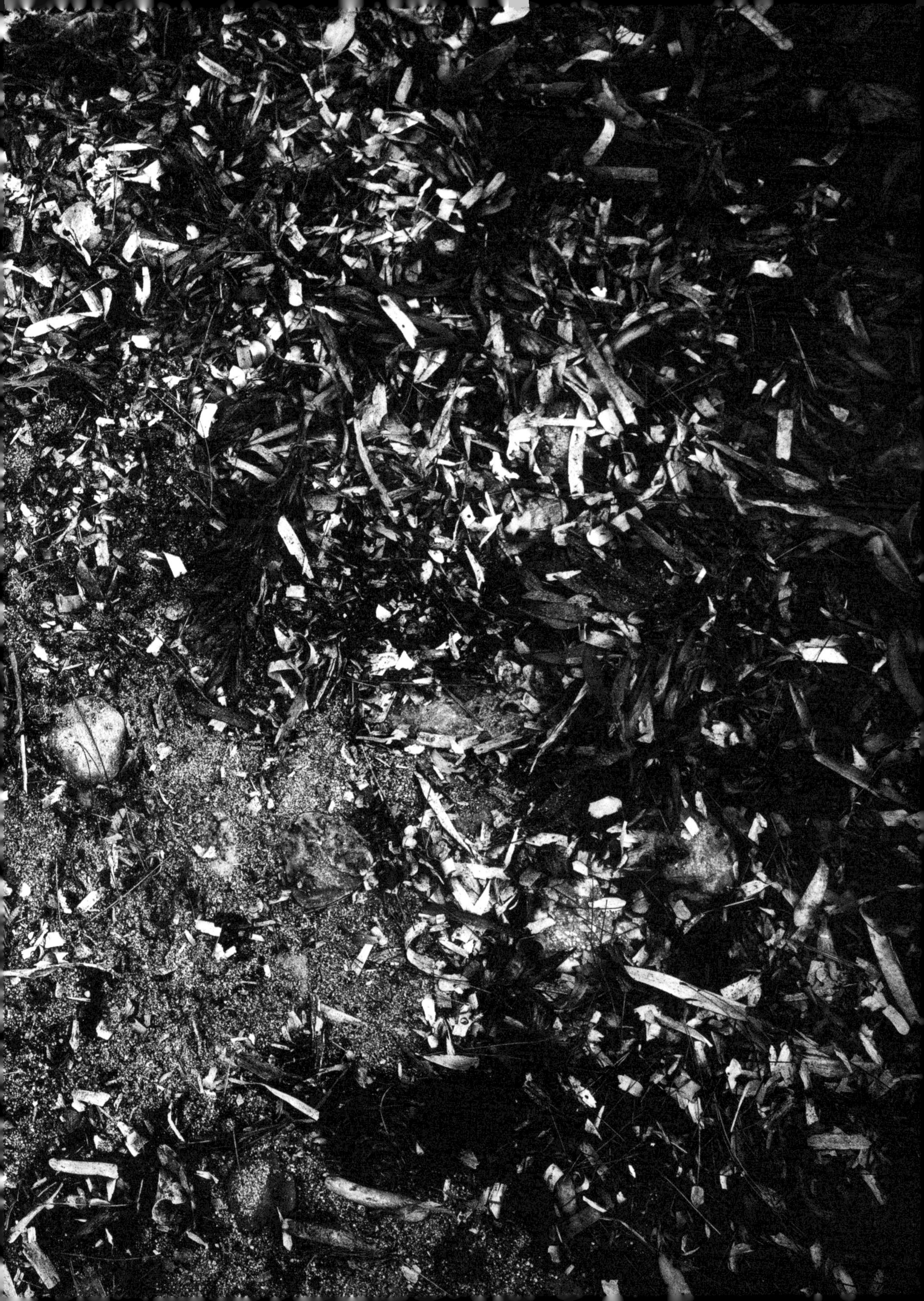

'We fell into the sea, shouting the children's names amid screaming voices and crying. We swam and counted heads [...] The children started asking again if we were going to die. Four hours passed like that. [...] "Let it finally end" the children were crying. But it did not end. At 22:00 a boat with divers arrived. We wholeheartedly thank these wonderful young men. They were volunteers from a diving school.' [8]

Elisavet Papadopoulou

'Make up a report about the fire, it is not the fault of one, or the other... Make up a conclusion with your experience...If there are incomplete data, you don't care...The Prosecutor is not going to tell you anything...' [9]
Vasileios Mathaiopoulos Head of the Fire Department (2018)—
Recorded Intimidation to the Forensic Expert D. Liotsios

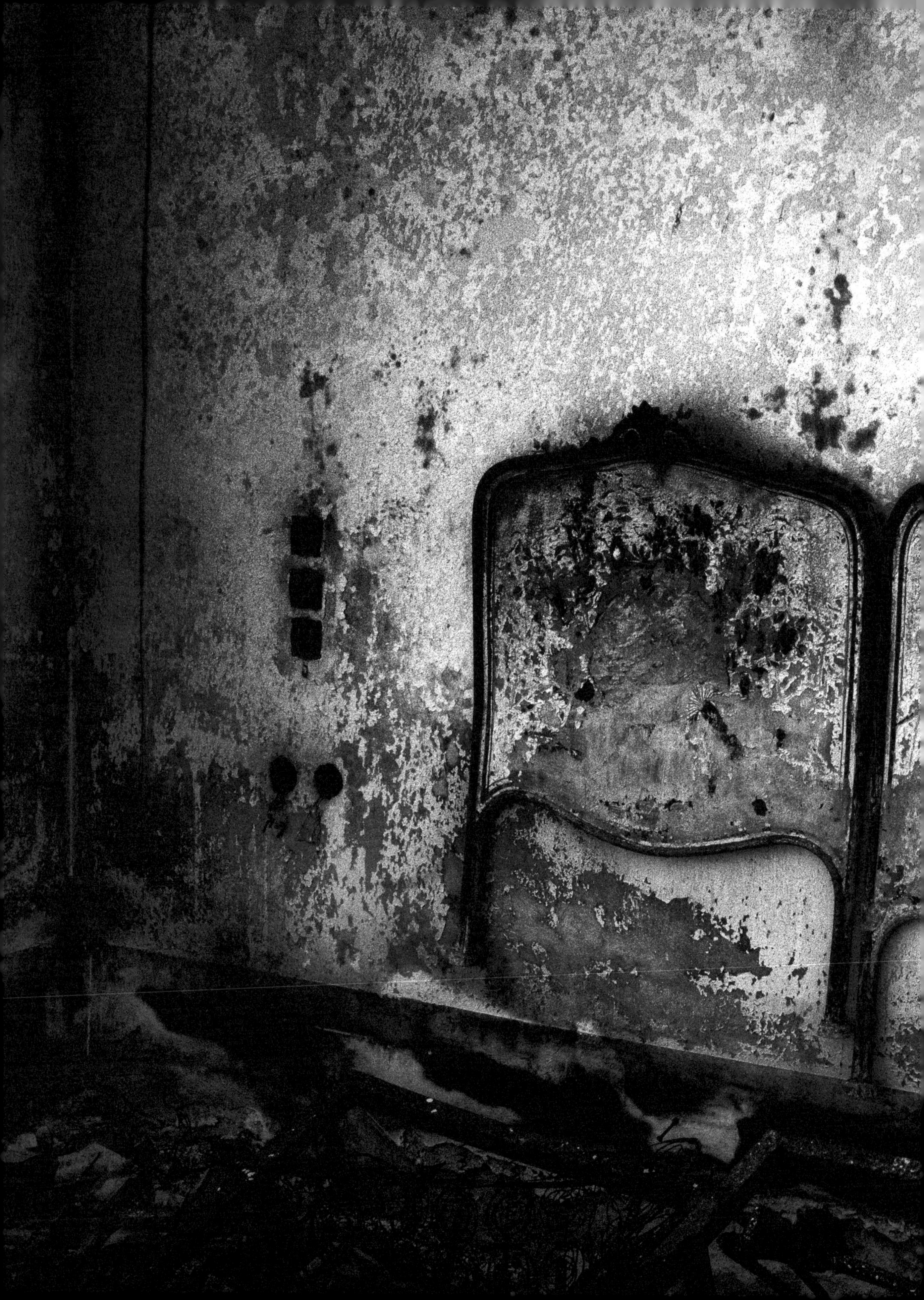

'When the flames approached we fell into the water. [...] It was before 19:00 in the afternoon, but we saw nothing but darkness. The water swept us away, we stayed at sea for many hours, five to six hours, until the currents took us out to Loutsa. At around 23:00 a trawler picked us up. We lost people in the meantime, a woman drowned as well as a young man. For a while I swam next to a floating corpse. She was a woman with white hair.' [4]

Nikos Stavrinidis

'If by 17:30 the ESKE (Crisis Management Command Center) had reacted immediately by sending three aerial firefighting aircrafts—which it had the ability to send—the fire would at least had been controlled, if not very likely extinguished...' [10]
From the Report of Athanasios Marneris, State Investigator

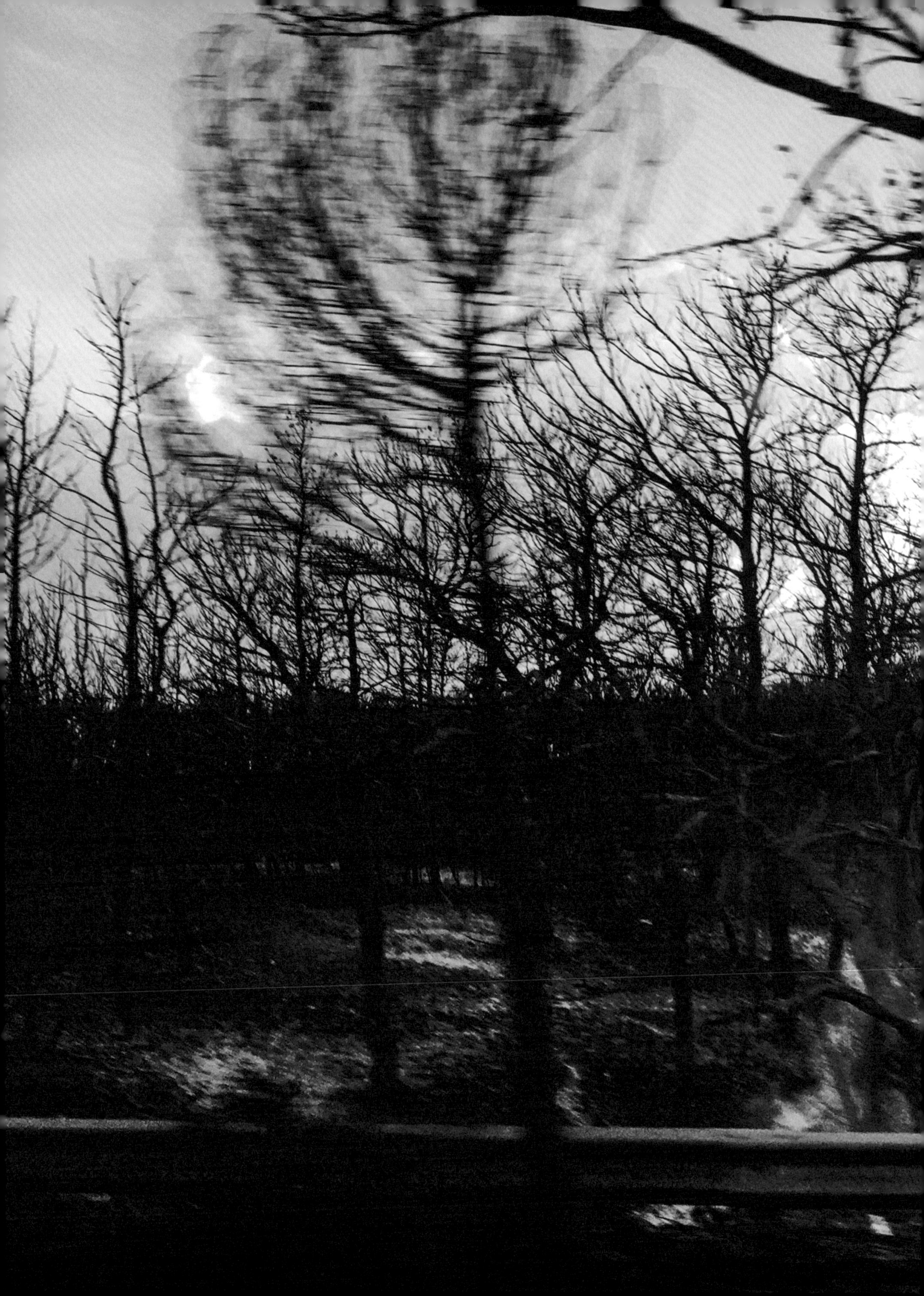

'...But our plot ends at a cliff; there isn't a beach below, but there are rocks, accessed by some rough steps that my grandfather had carved into the cliff face. We've had the plot since the 30s. The first group, the people we saw, made it. Others behind them didn't...' [11]
Christianna

'Which sea? It was hell there. We entered the water at 18:25. We had no intention of swimming, nor of reaching the opposite side to Evia. [...] We started hearing explosions. The sea got wilder. A black veil. Everything got dark. The sea was very rough. Too much wind, too many waves, too much smoke. [...] We stayed there for about an hour, struggling to keep him out of the water. He couldn't take it anymore, he vomited. He raised his hands, asked God for forgiveness, thanked us, "thank you for everything you've done for me," he said, and took his last breath [...]' [12]
Resident

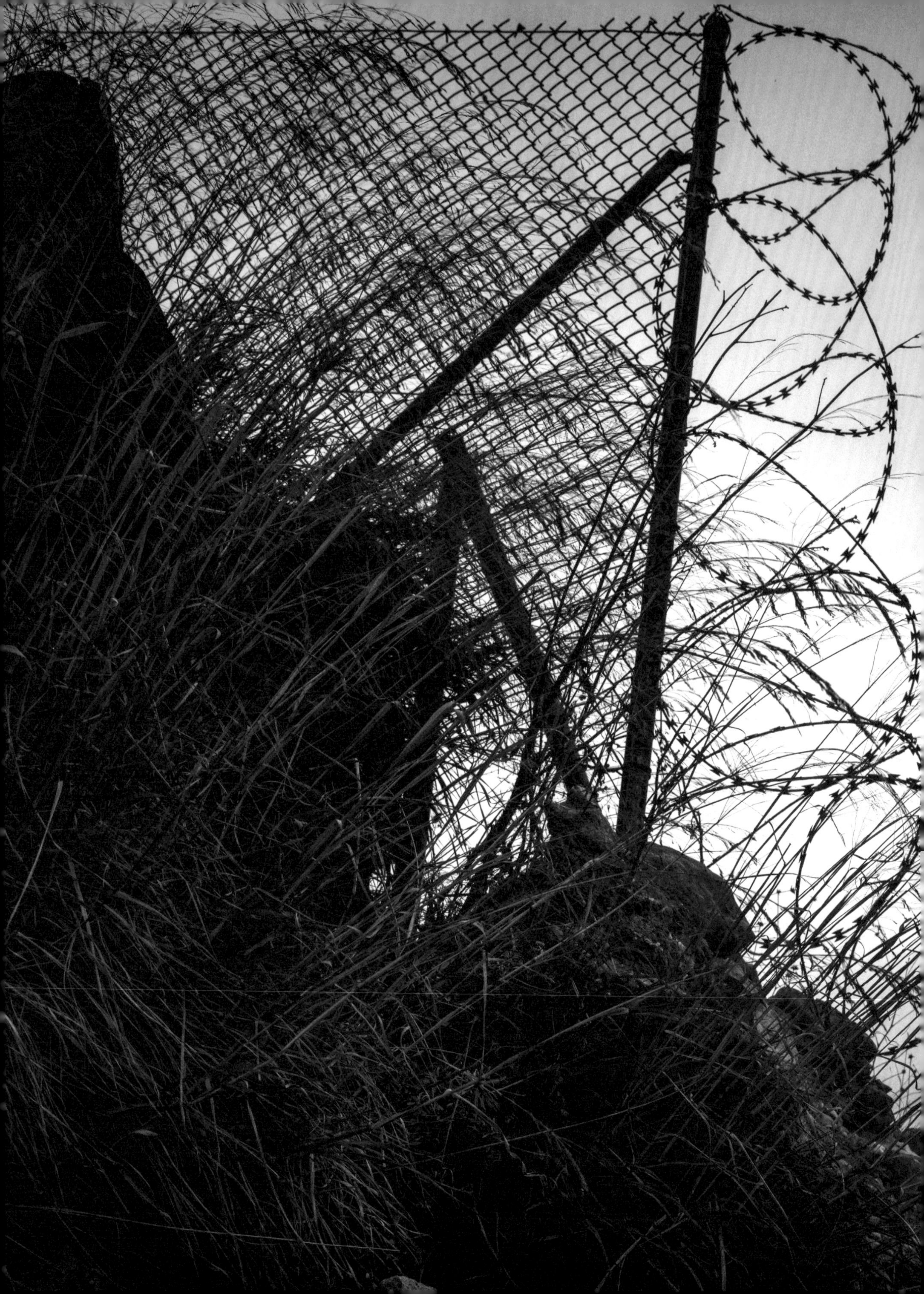

'It was wrong coordination. They sent everyone from the Marathonos road to take the coastal road because they did not believe that the fire would end down by the sea. This had never happened before. Everyone ran in panic and got stuck here.' [13]

Giannis Stampelos

‘Everything the state had to do, the people did from the first moment. [...] I may have been able to get back on my feet, but I have not slept for more than three hours a night for months. I cannot forget what I experienced.’ [14]

Dimitris Filippis

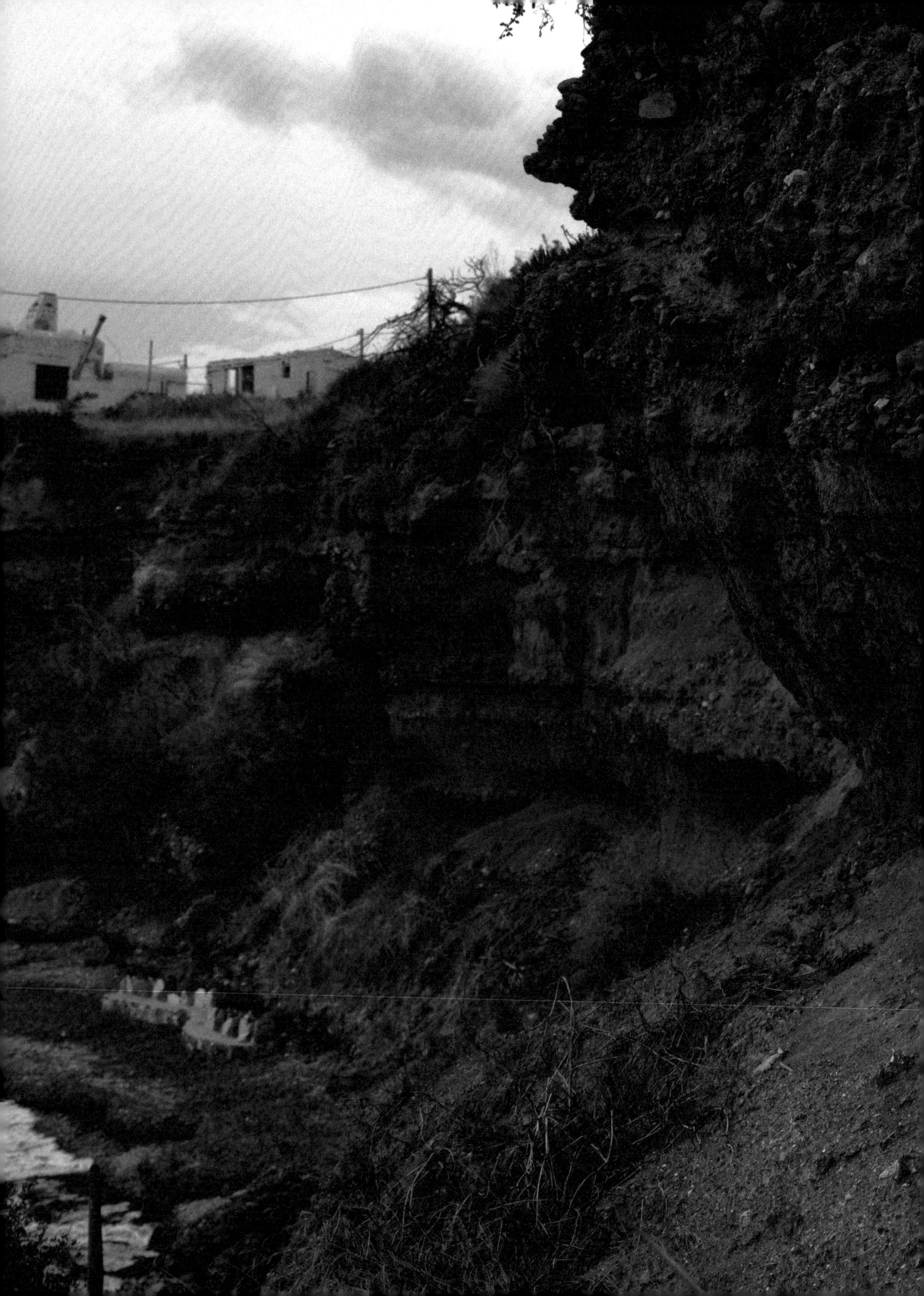

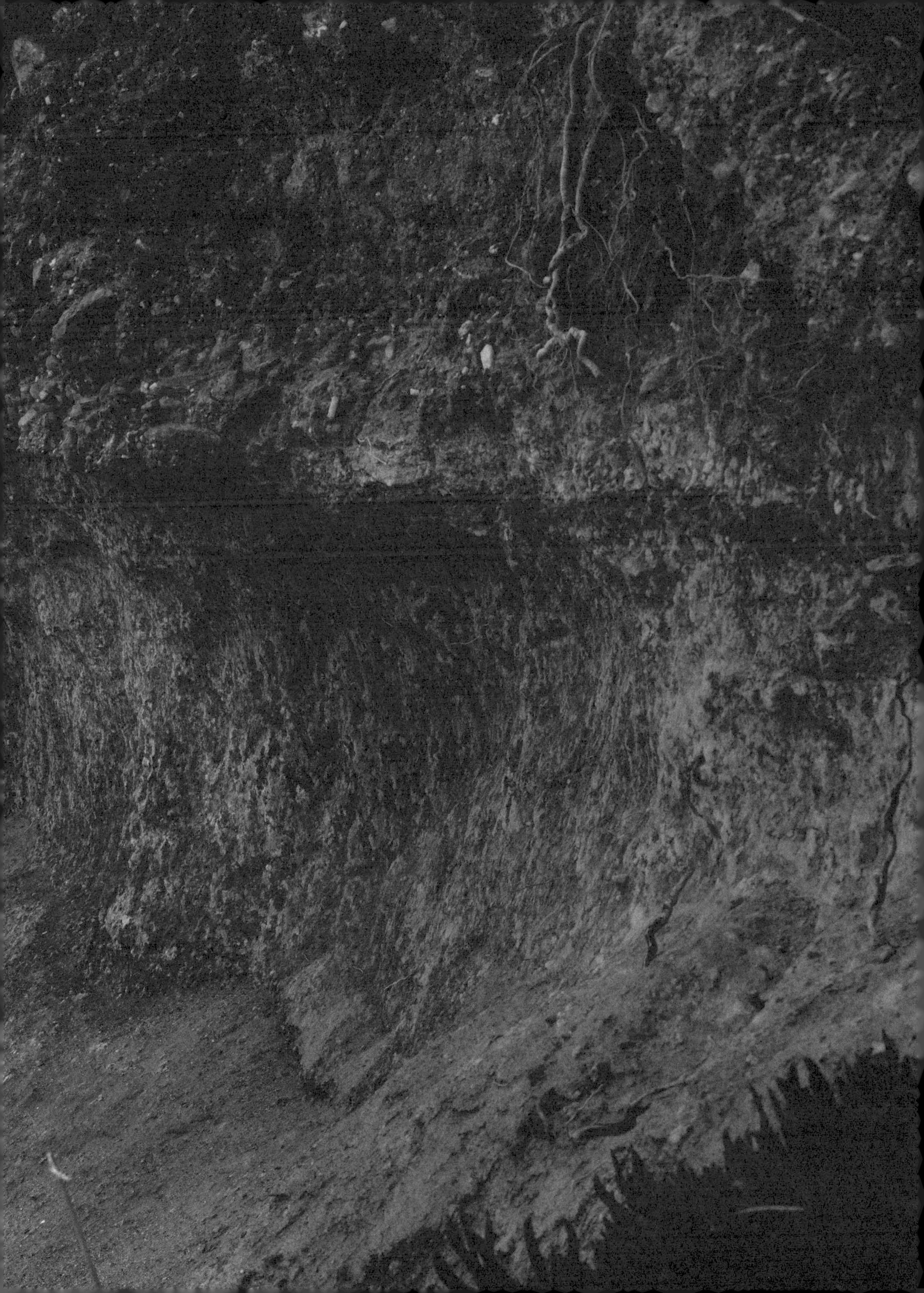

'We were the first boat to approach the beach in Mati. The landscape was nightmarish, bizarre, the smoke had covered the surface of the sea, I had never seen this before. Flames almost to the water, explosions, voices of agony, what can I tell you? [...] After hours of effort we managed to rescue a total of seventy-three people...Some had burns, others were screaming or crying. I pulled many of them out of the water myself with one hand—I do not know where I found such strength.' [15]
Mahmut

'They (the government agencies responsible) were observing uninvolved not unlike Xerxes when he was watching the battle of Salamina from his Golden throne on Mount Egaleo.' [10]

From the Report of Athanasios Marneris, State Investigator

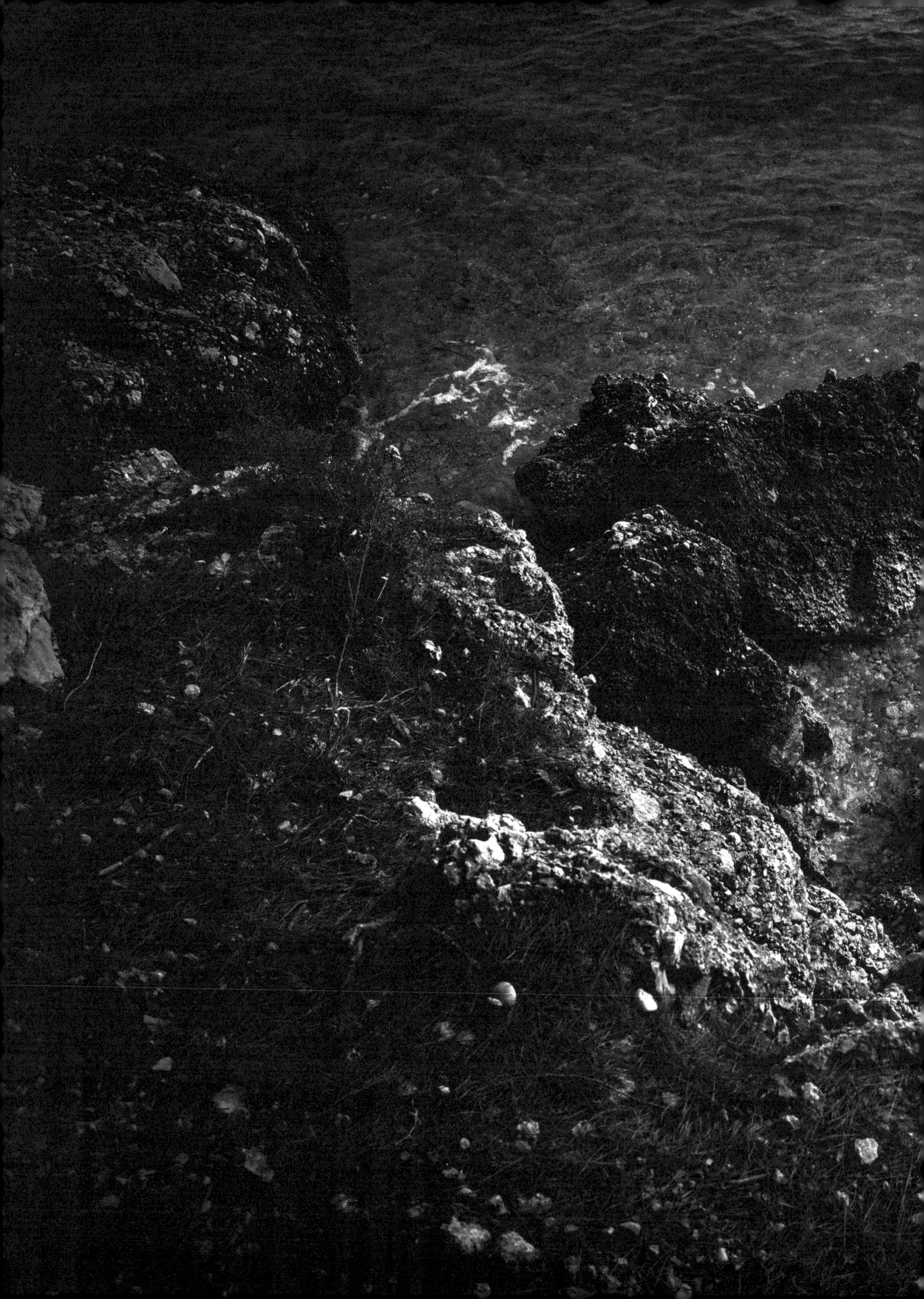

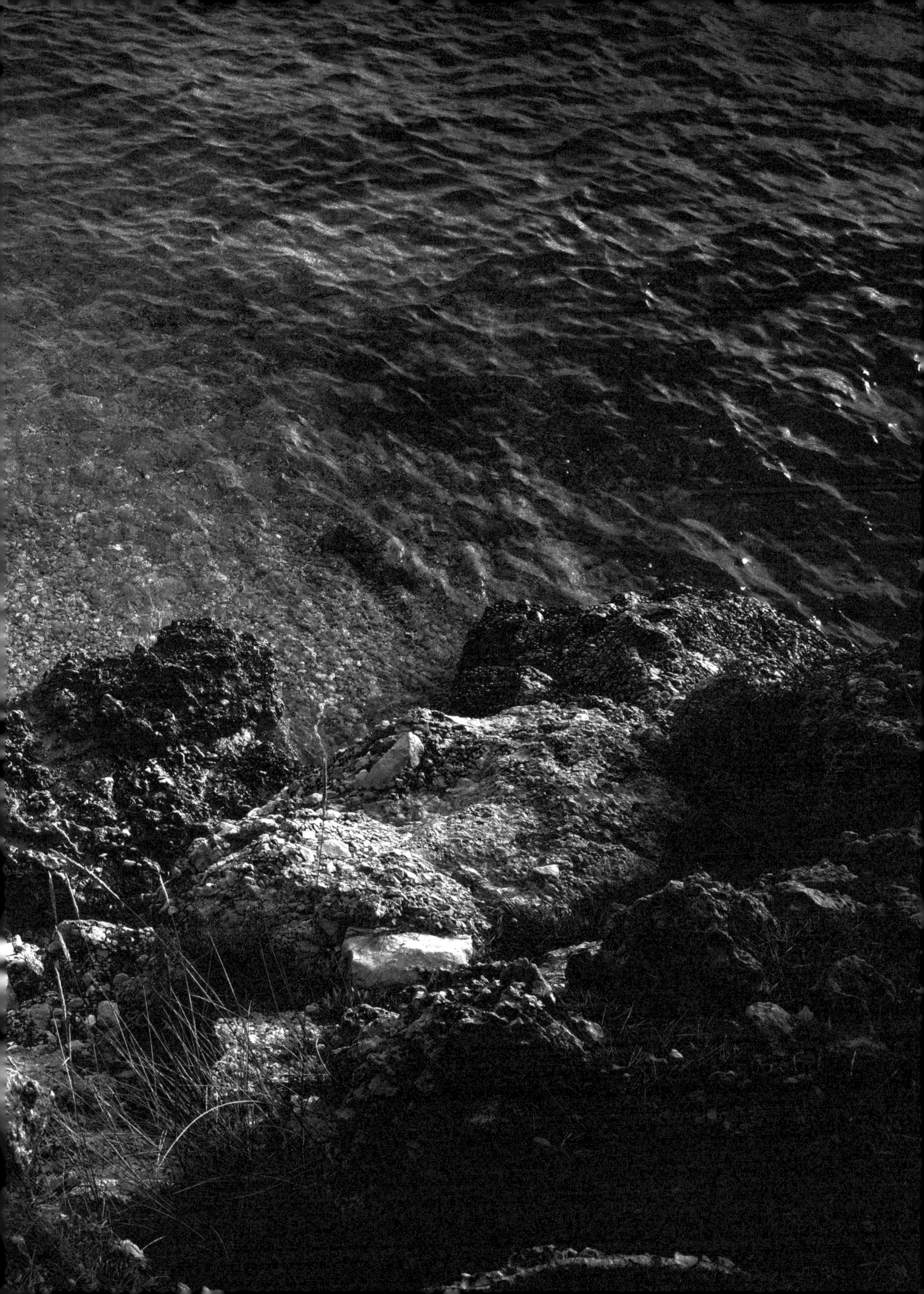

'What are you doing showing off your skills with the beasts, with 100 dead. "Where are you going? If you write about the responsibilities of your superiors, we will all gather and tear you apart."' [16]

Vasileios Mathaiopoulos Head of the Fire Department (2018)—
Recorded Intimidation to the Forensic Expert D. Liotsios

'Mommy, I want to sleep forever...' [17]
Young child of Agni

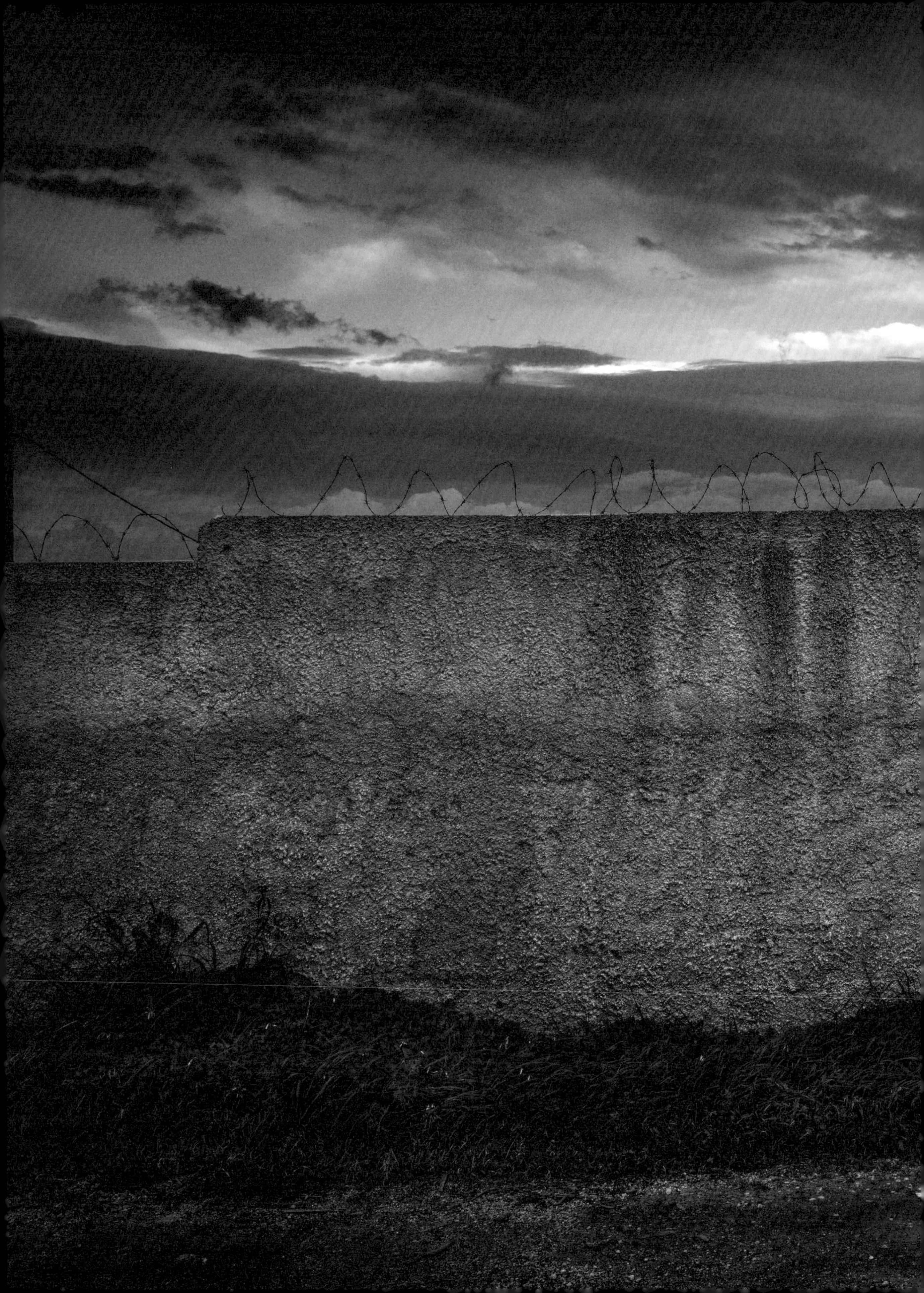

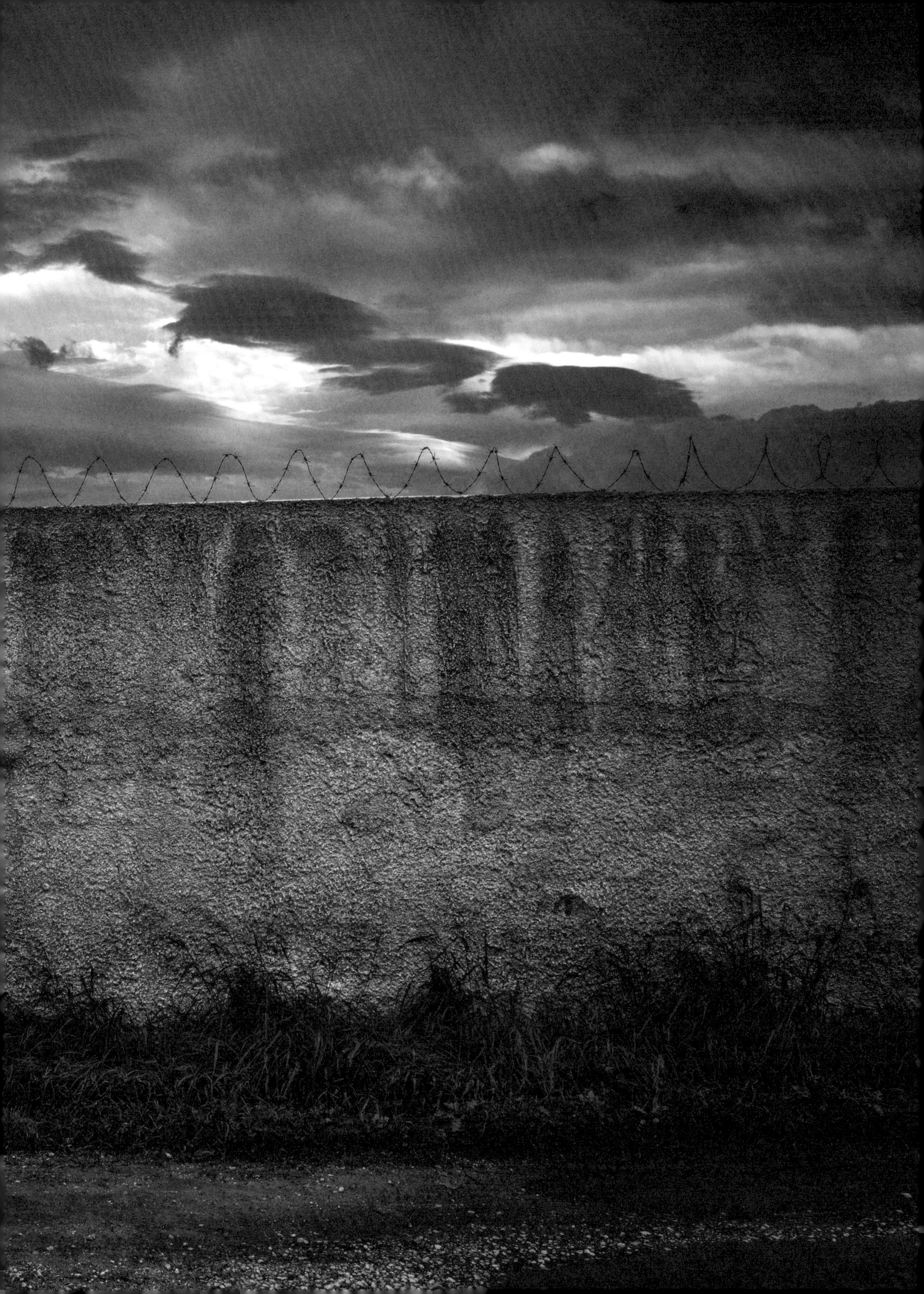

'I called 112, the emergency telephone number. They said they would come to our rescue, but never did. I still haven't forgiven myself for not putting Margarita and the baby over my shoulders and doing more to get to a hospital. When we finally got them to an ambulance, it was [too] late. The baby died later that night, and Margarita a week later.' [5]
Menelaos

'"Does the prosecutor know what papers you have asked for?" Mathaiopoulos asks. Liotsios explains that most of the documents have already been notified to the prosecutor's office. Then the former head of the Fire Department urges him to see which ones have not been notified and to bury them: "Look who has them, who does not have them...and bury them. Calmly, no need to panic..."' [16]

Vasileios Mathaiopoulos Head of the Fire Department (2018)—
Recorded Intimidation to the Forensic Expert D. Liotsios

'For the rest of the world, the fire belongs to two years back. For us, however, it still burns, it still smells. It is difficult to walk some streets knowing that your friends were lost there.' [18]
Aphrodite Hatzianastasiadis

'This is what drives me crazy [...]. You go through a disaster like this and you have no help, but extra burdens. "It's your fault..." they tell you. This burden will never go away.' [18]
Dioni Toupsi

'There are at least twenty-five pathways to the sea. From Marathonos Avenue to the sea it is 850 meters. The only tricky area is at Kokkino Limanaki. Throughout the rest of the area all roads lead to the sea.' [19]
Dimitris Liotsios, Fire Department Forensic Expert

'It is a mass disaster with a lot of carbonized bodies, which are hard to identify. We expect the identification with DNA methods. We are trying to determine the cause of death, that in many cases is obvious.' [20]
Nikos Kalogrias, Forensic Pathologist, Athens Forensic Service

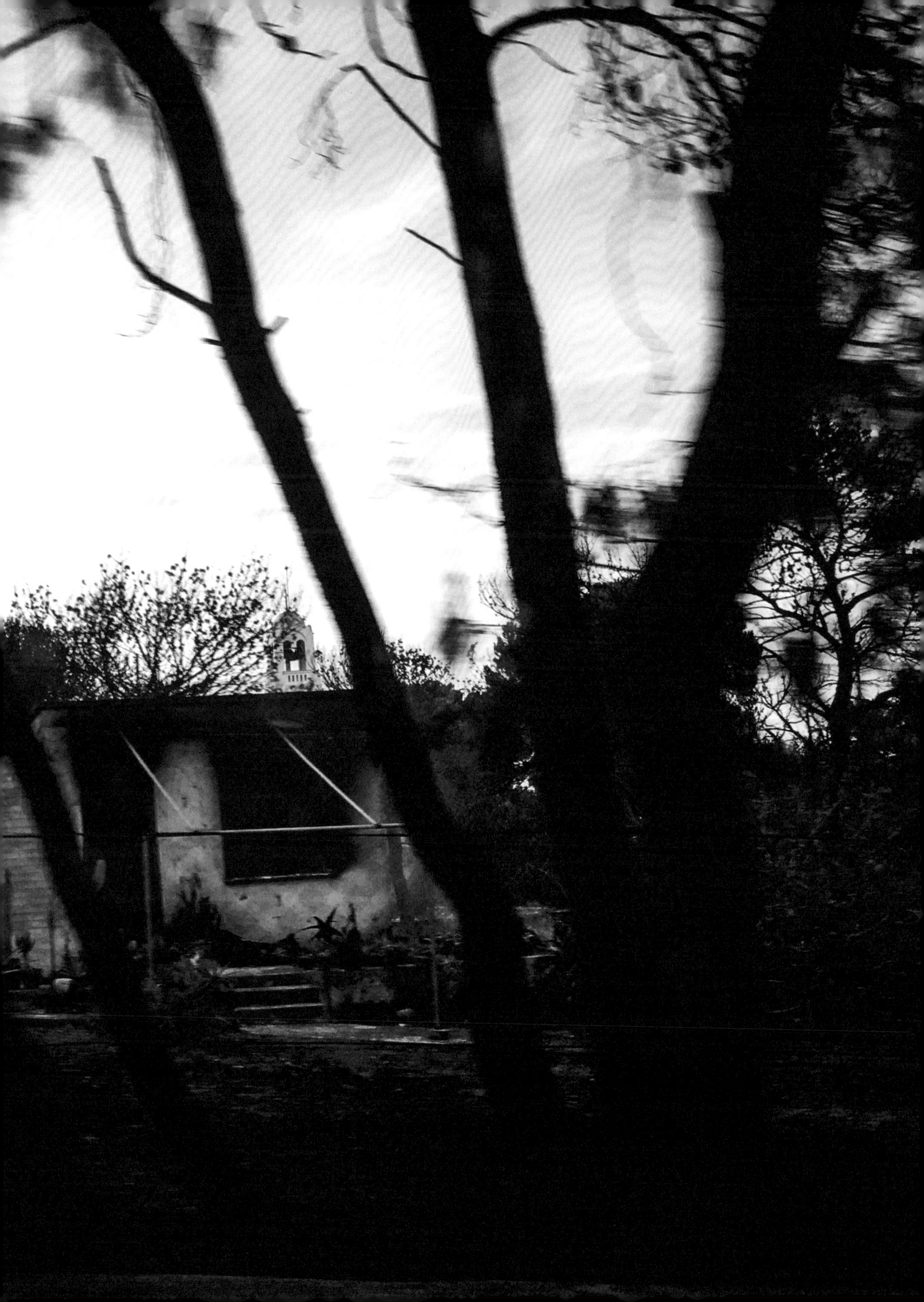

'They (my family) left at 18:30 and headed towards the hill. Shortly after, they were burning alive along with a couple who were trying to escape. At the time that my family and the couple had caught fire, it was the first time that the Civil Protection Operations Center of the Fire Department contacted the Municipality of Marathon. That's how late after the fire started.' [21]

Akis Herouveim

‘The fire in Kineta will cost 500 million, the fire in Mati will cost 1 billion, of these 100 will be shared, 200 will be shared... Watch out... watch out...’ [19]
Vasileios Mathaiopoulos Head of the Fire Department (2018)—
Recorded Intimidation to the Forensic Expert D. Liotsios.

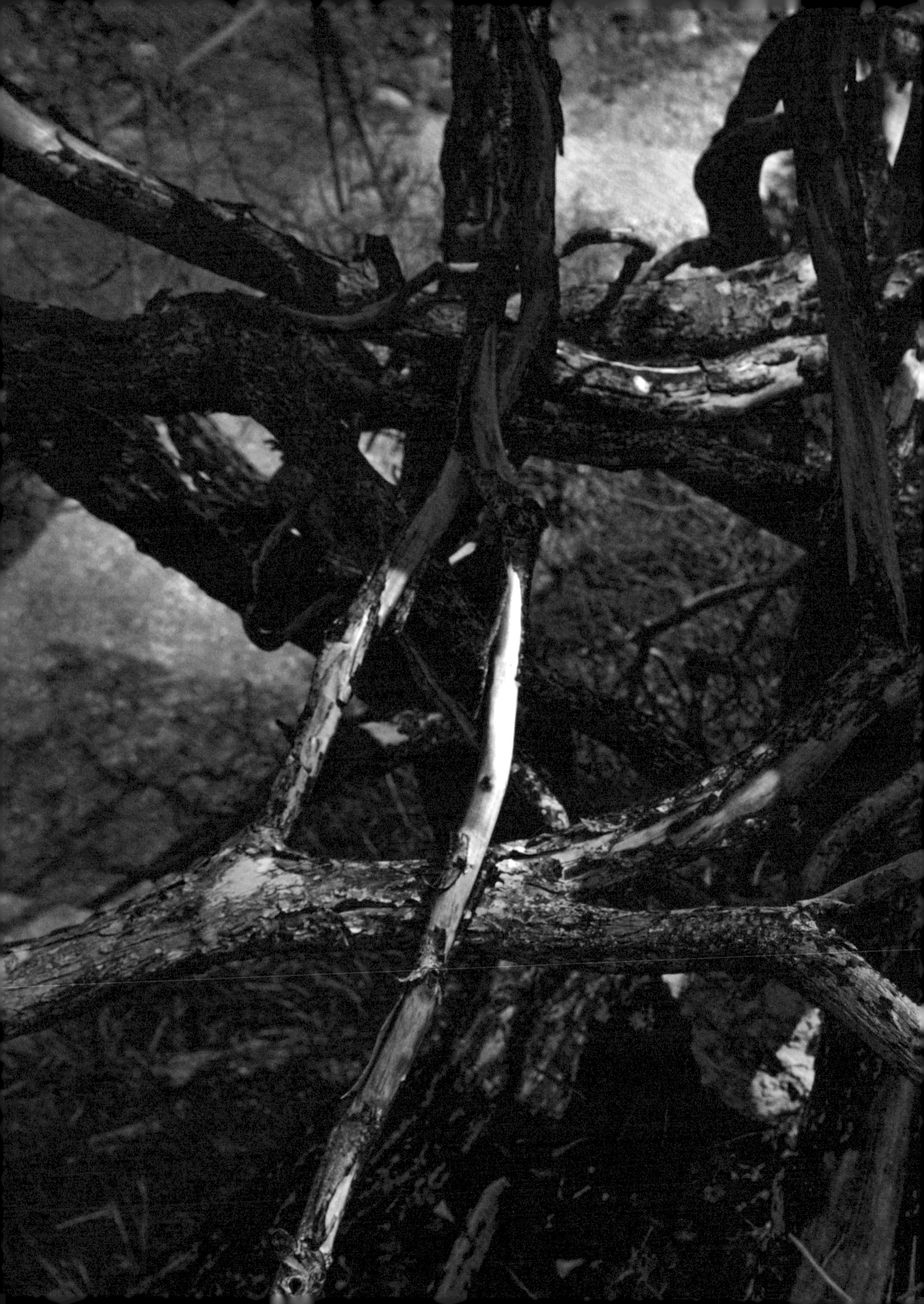

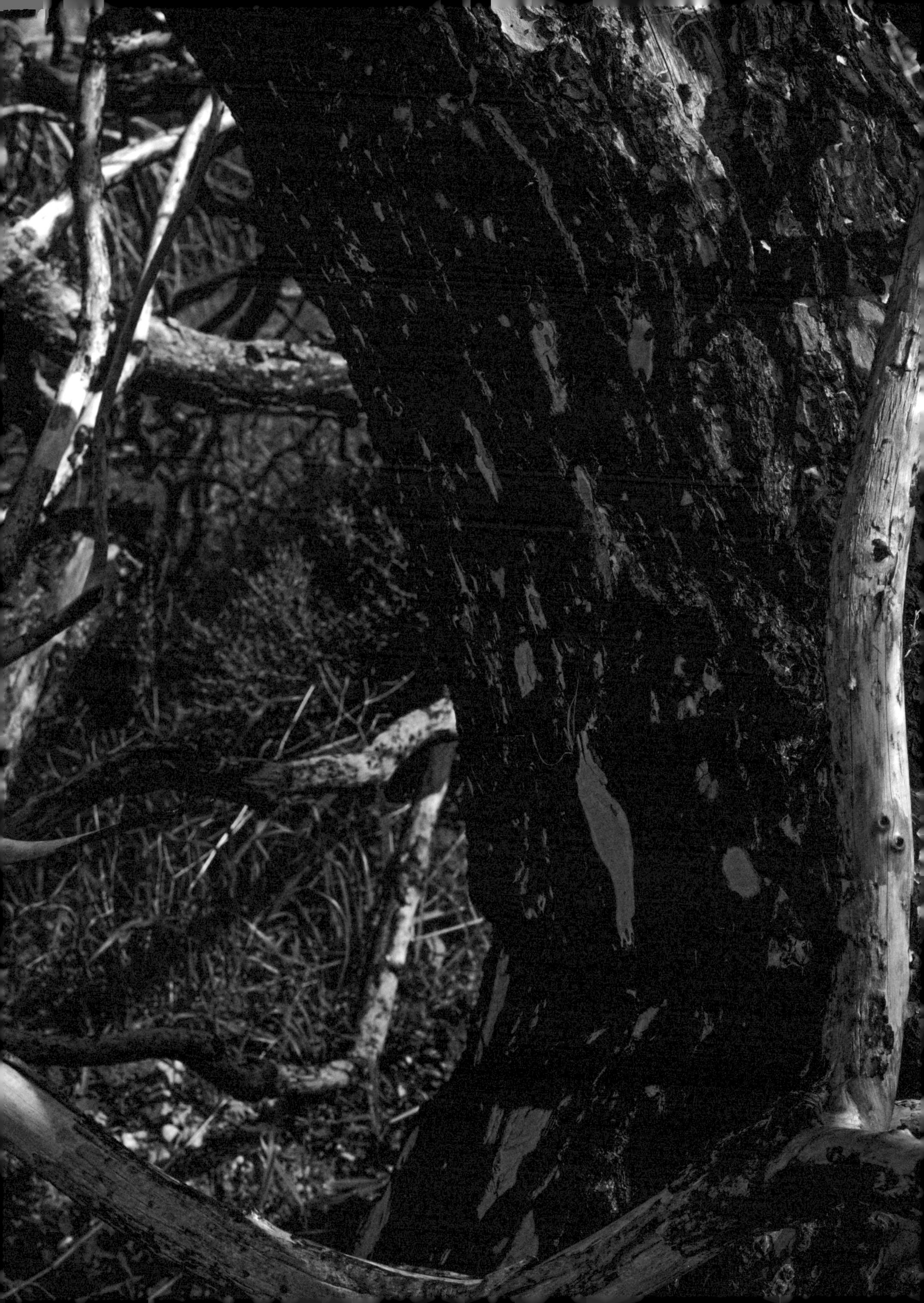

'My mother was lying on the sidewalk, down on her knees, so burned that she had no hair. She was naked, hairless and screaming. They brought her a tablecloth from the taverna. I know from a gentleman who was with her that she was alive for 5–6 hours. As soon as she arrived in the port in Rafina, she died.' [22]

Ioanna Petala

‘No one was shouting anymore—We were tired, you could only hear names. There were cries, groans, we were tired, there was a muffled groan of humans that couldn't breathe and we kept hearing names being called. [...] Why did they forget us? Why did no one come? Why?’ [17]

Agni Gkantona

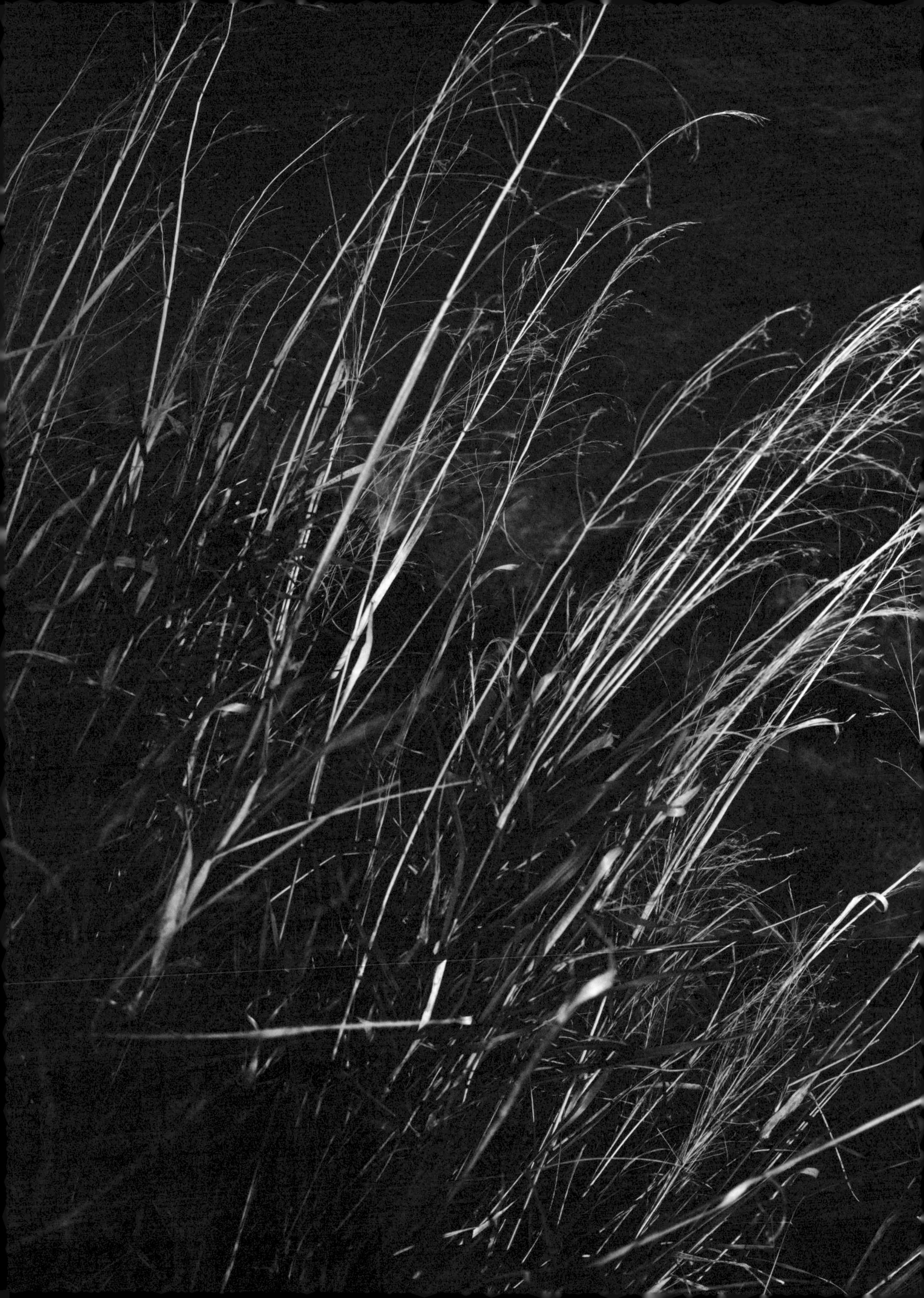

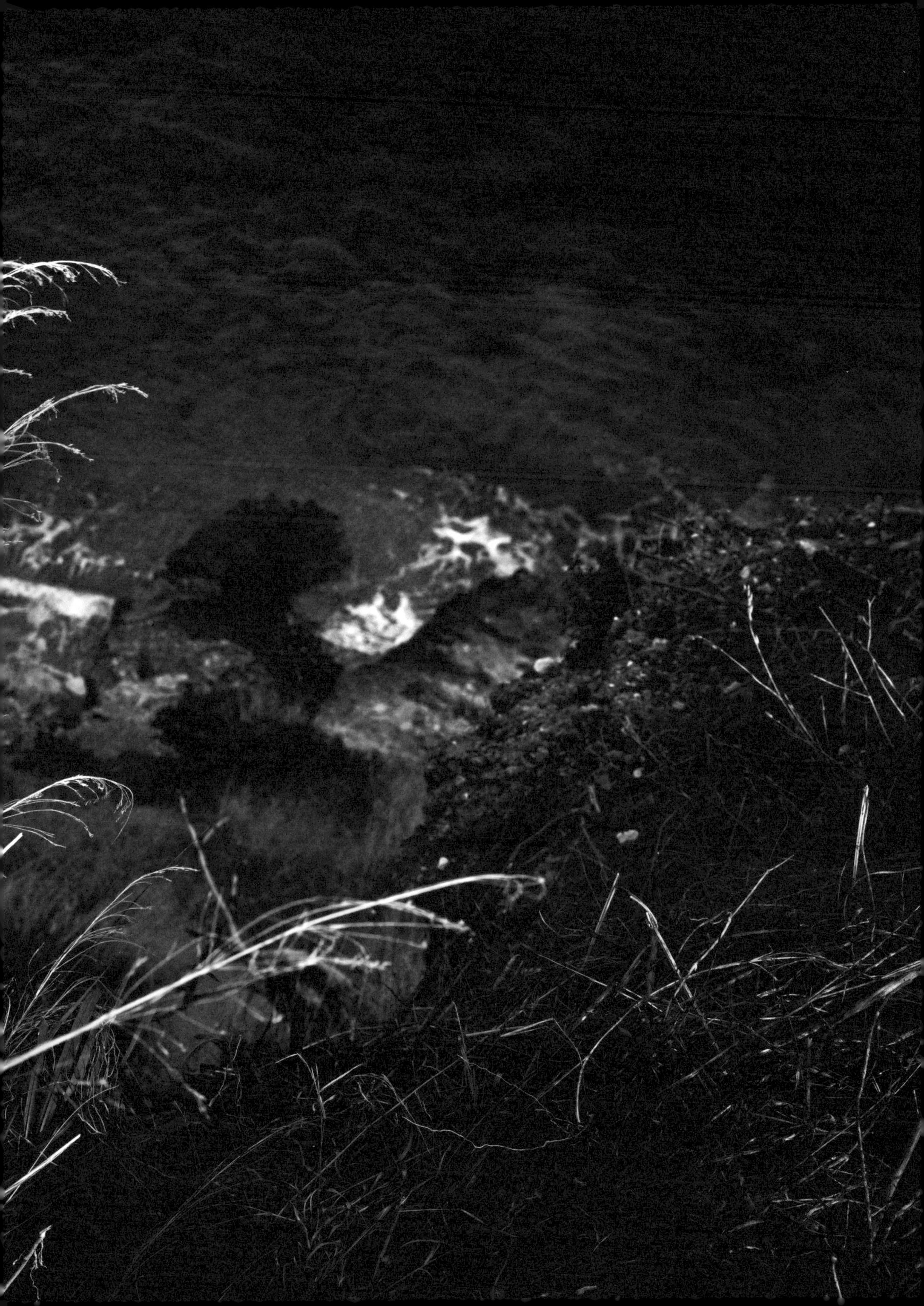

'I am an old man. I might not be able to talk about my child again, I might not have enough time left. She died 140 steps from the sea. I count them every year; at the memorial service. Do you understand what I am saying?' [21]
Aggelos Siakparas

'[...] from Marathonos Avenue to the sea, it is 500–600 metres, which means that the average resident of the area could be in a safe place near the sea within fifteen minutes at the latest. Therefore, there was time and sufficient forces in order for the evacuation to take place in a safe way. We were ready, however no order was given by the Fire Department.' [23]

Accused police officer

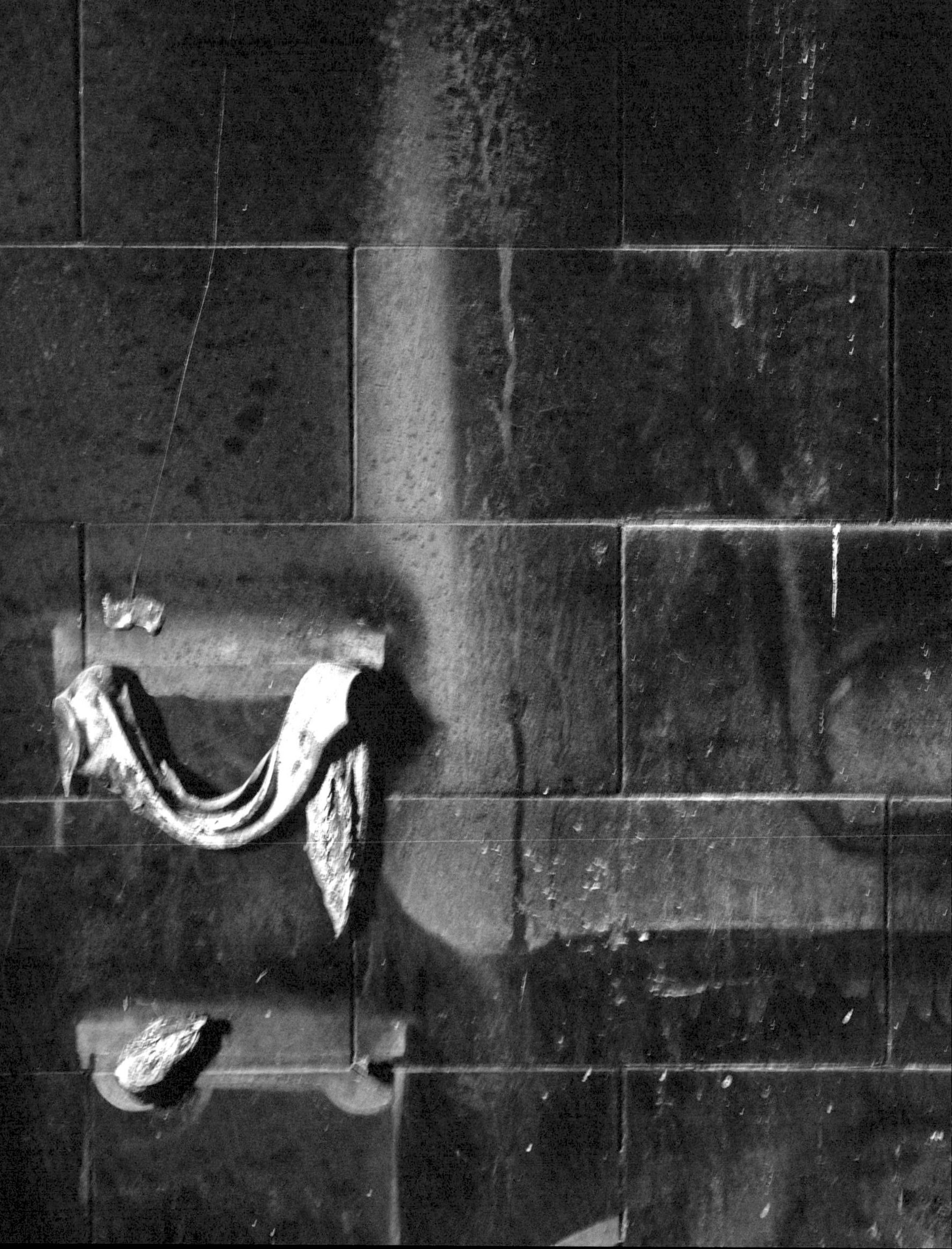

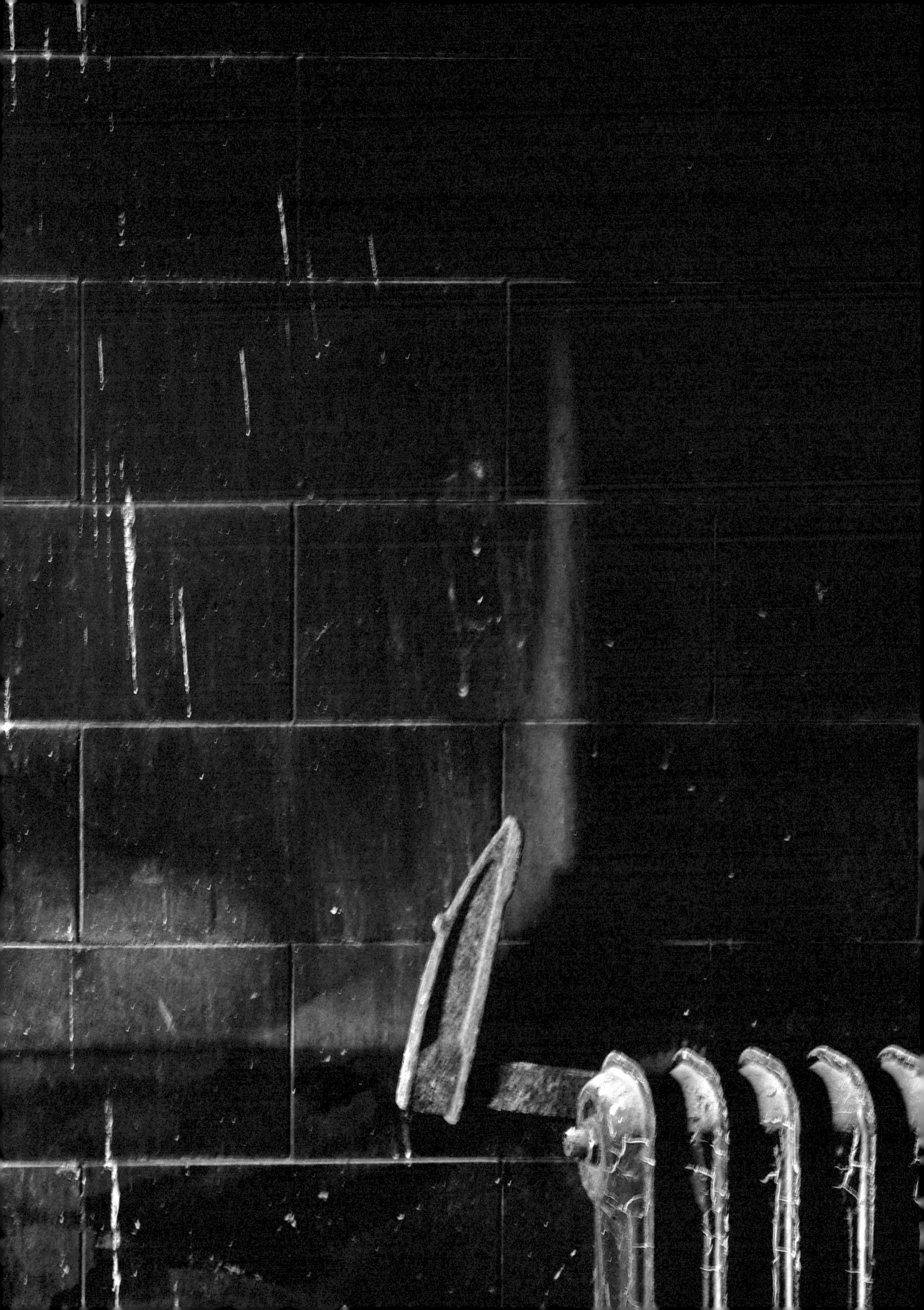

'I told my brother to let me die, but he wanted to give me another chance and I lived. I stayed in the hospital for eight months, six of them sedated in intensive care.' [24]

Mrs. Konstantaki

'I do not see big operational mistakes.' [25]
Nikos Toskas, Minister of Internal Affairs

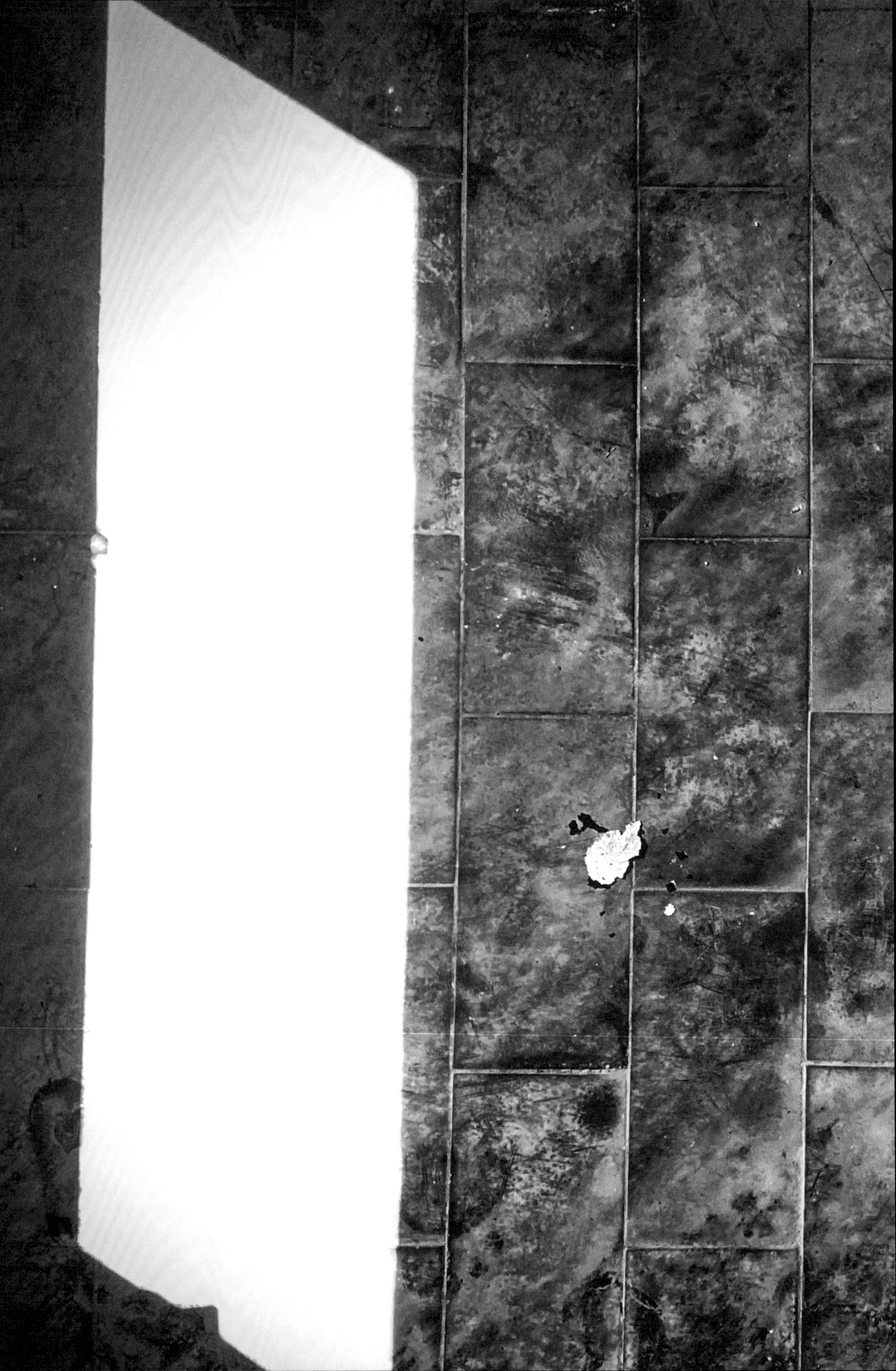

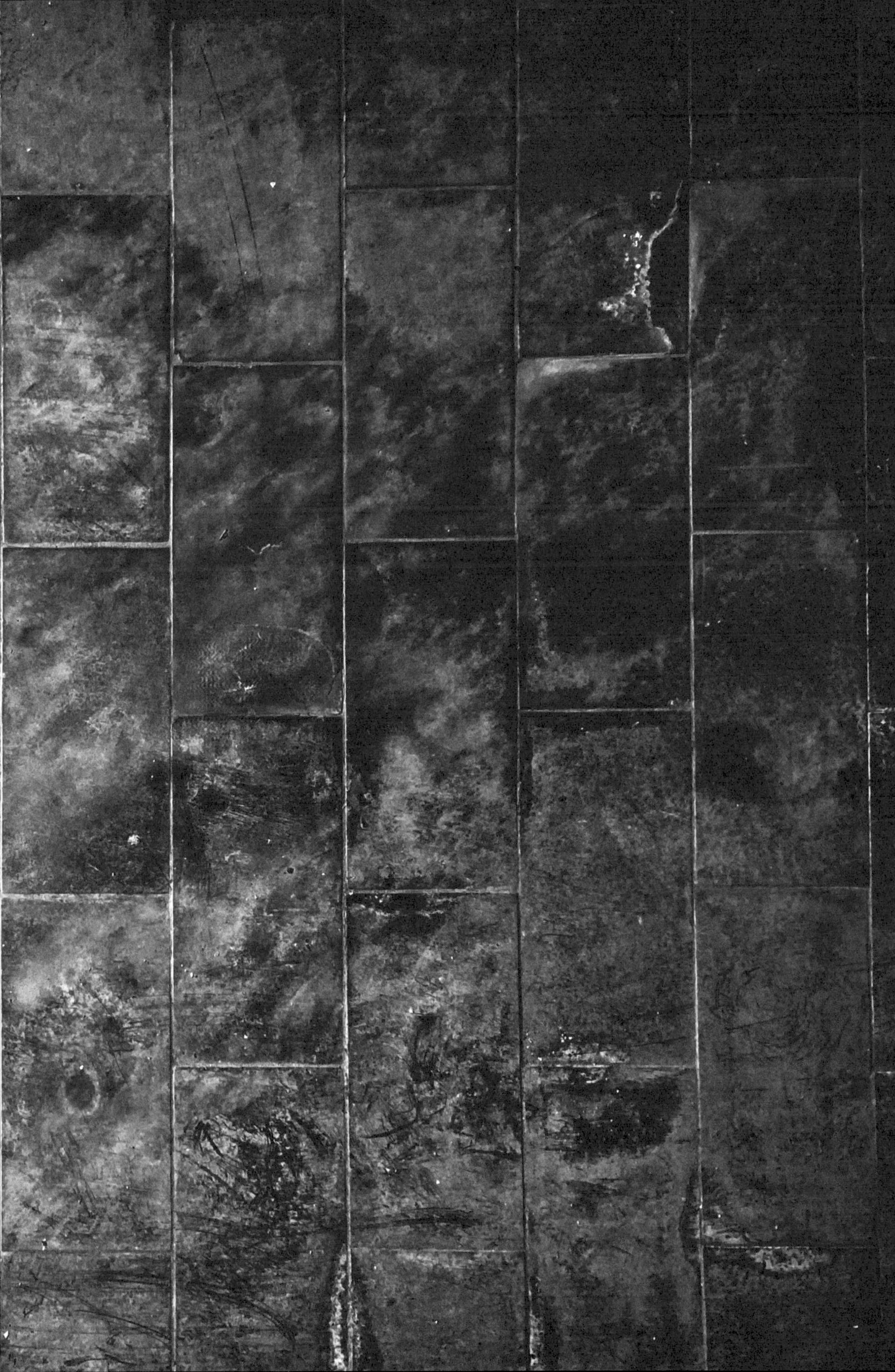

'I apologise to the families of the victims that I talk about us that lived.' [17]
Agni Gkantona

'During the great fires of 1997 in Penteli, a firefighter entered the yard and instructed us, "leave the gates open, hoses visible, and leave." In Mati—nothing.' [26]
Maria Tournaviti

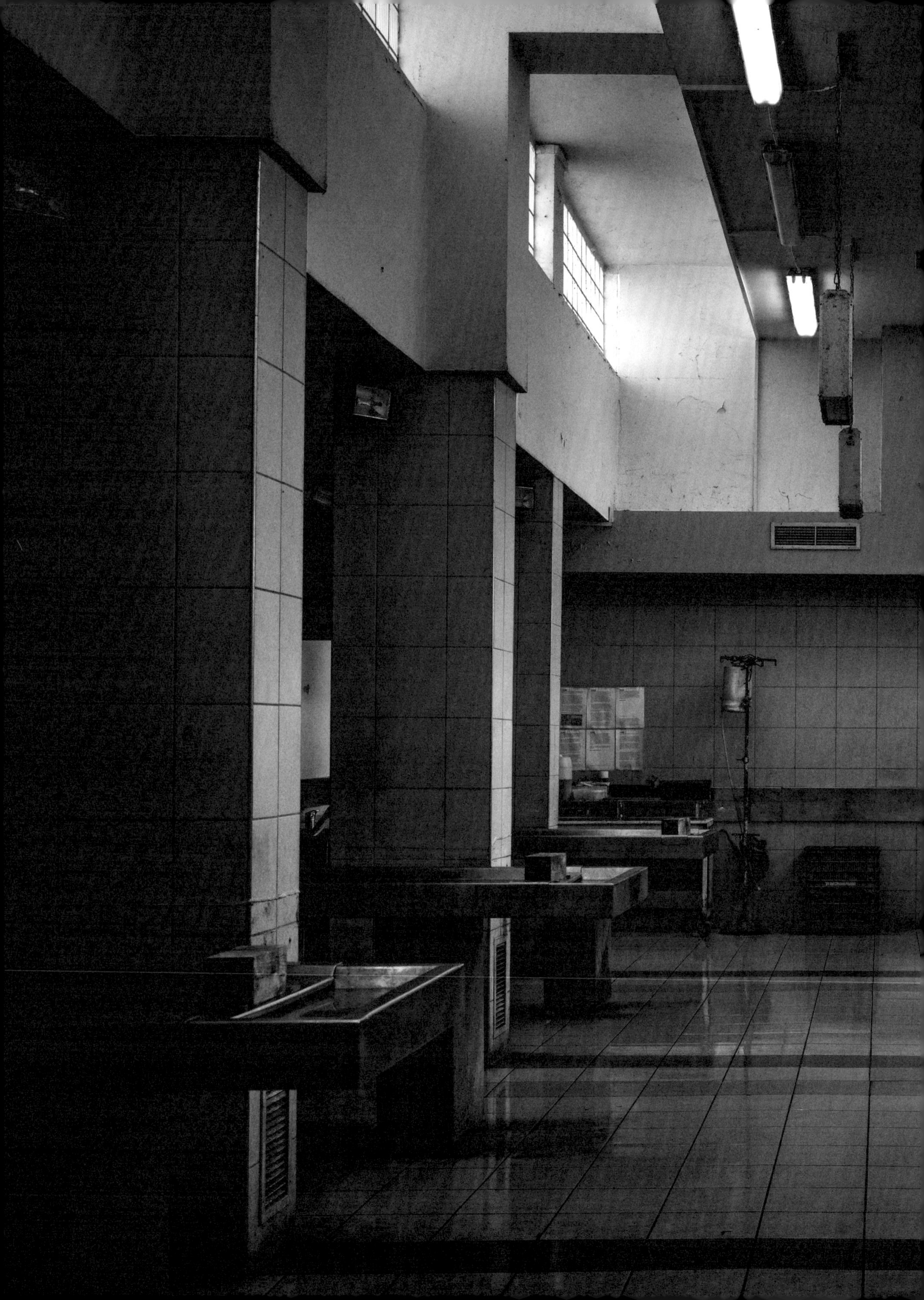

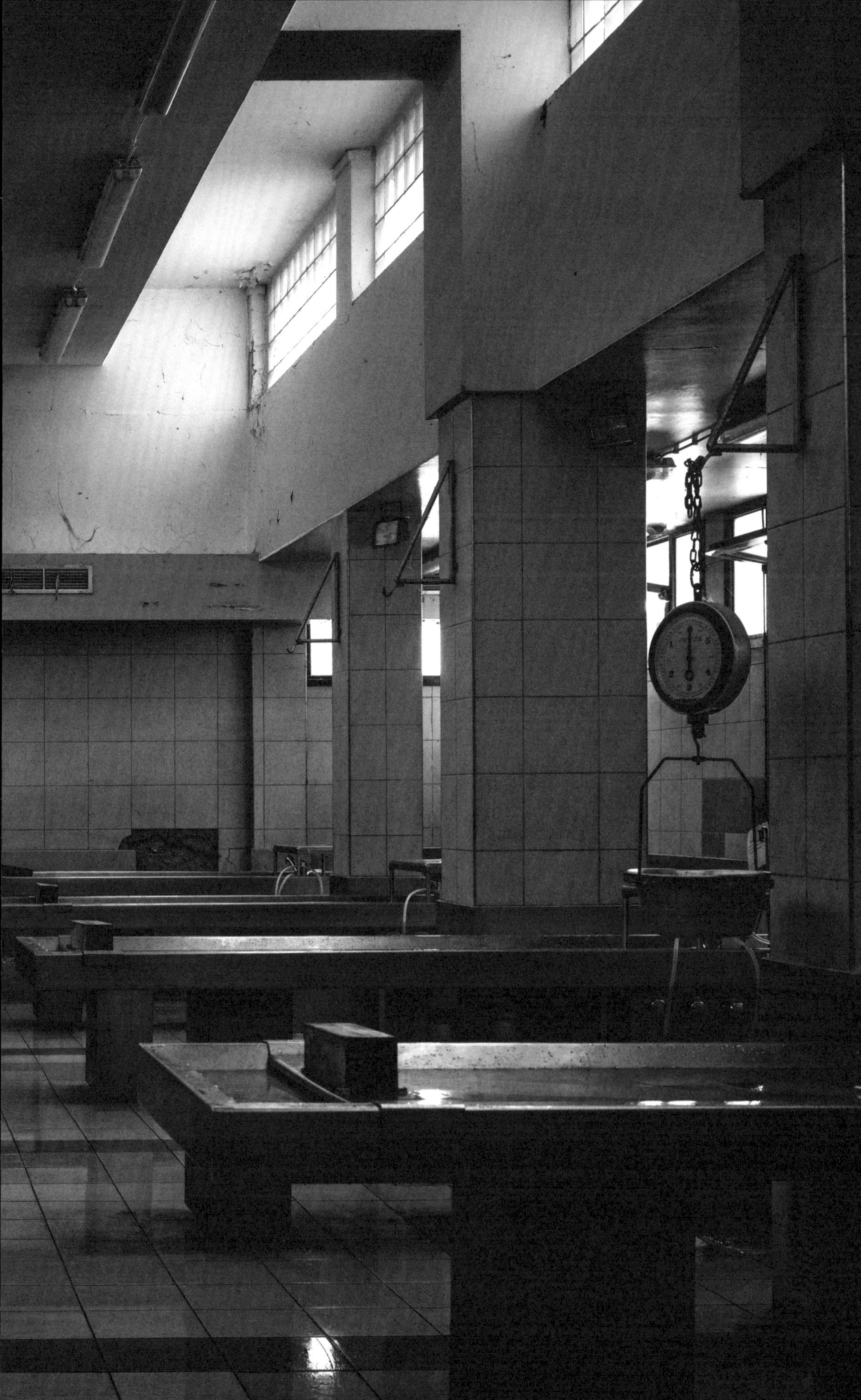

'It's not only what I saw, what I heard is worse.' [27]
Giorgios Potamianos

'I wonder, did I do the right thing? Could I have done more to save these two women, and I didn't?' [5]
Iakovos

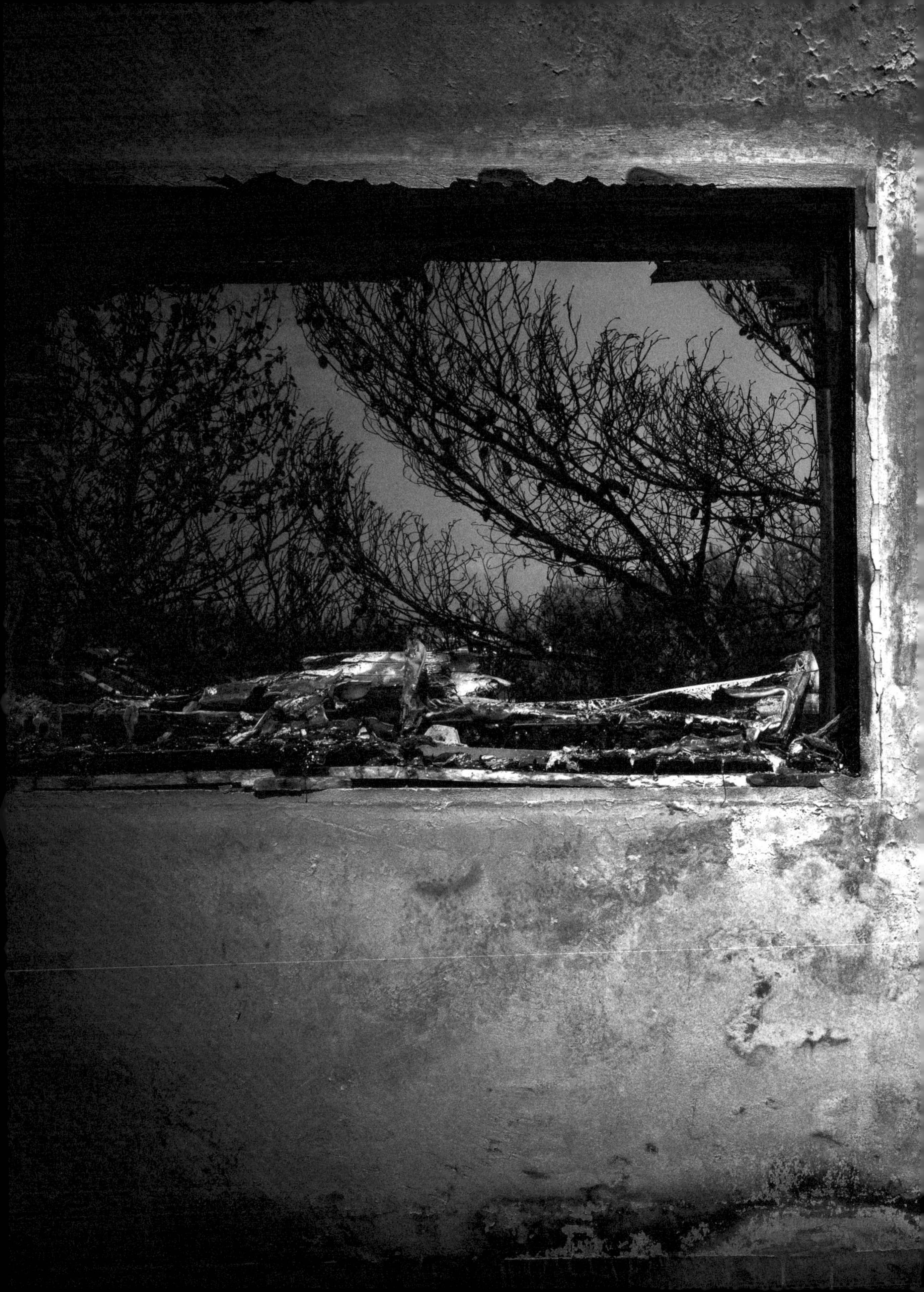

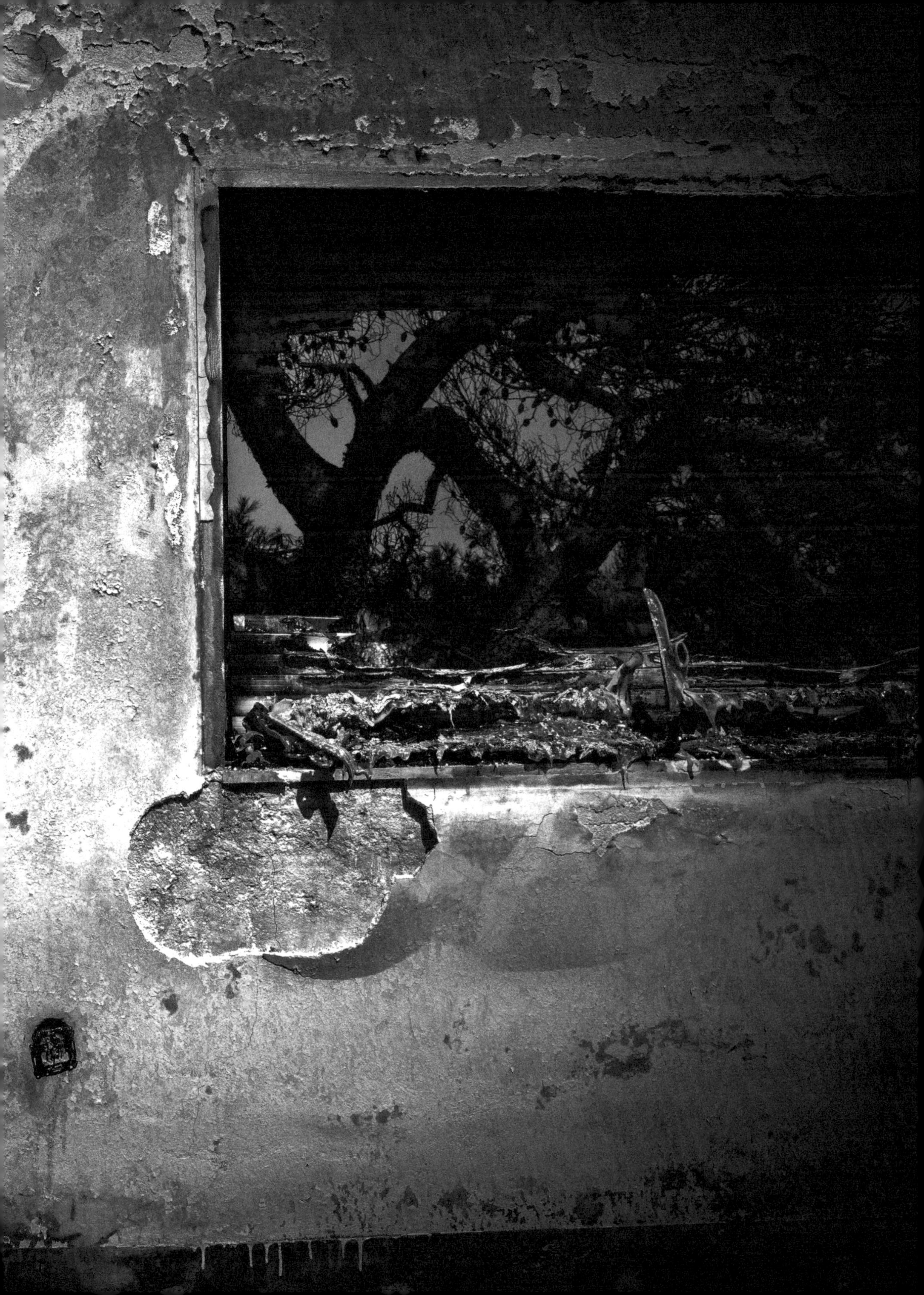

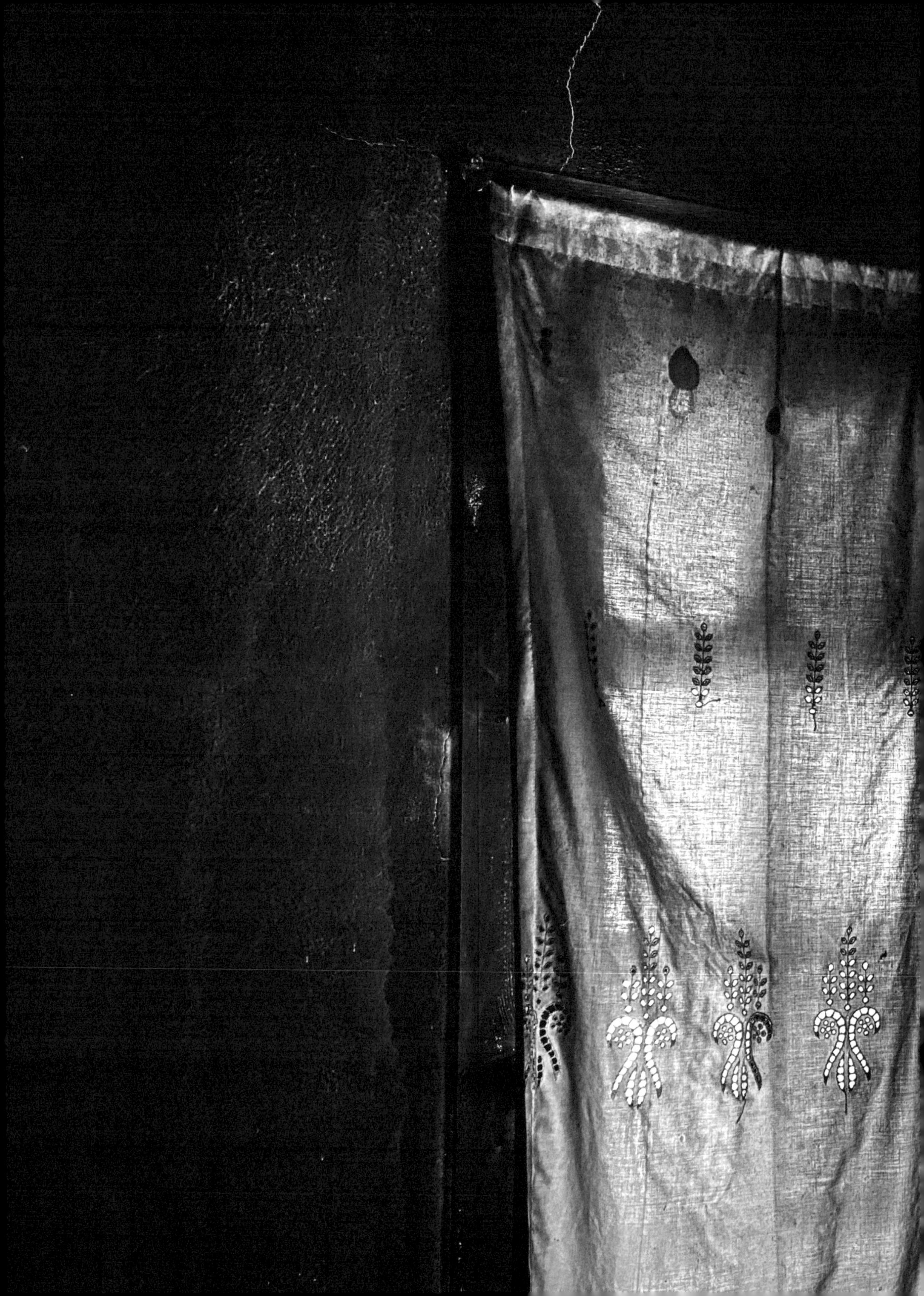

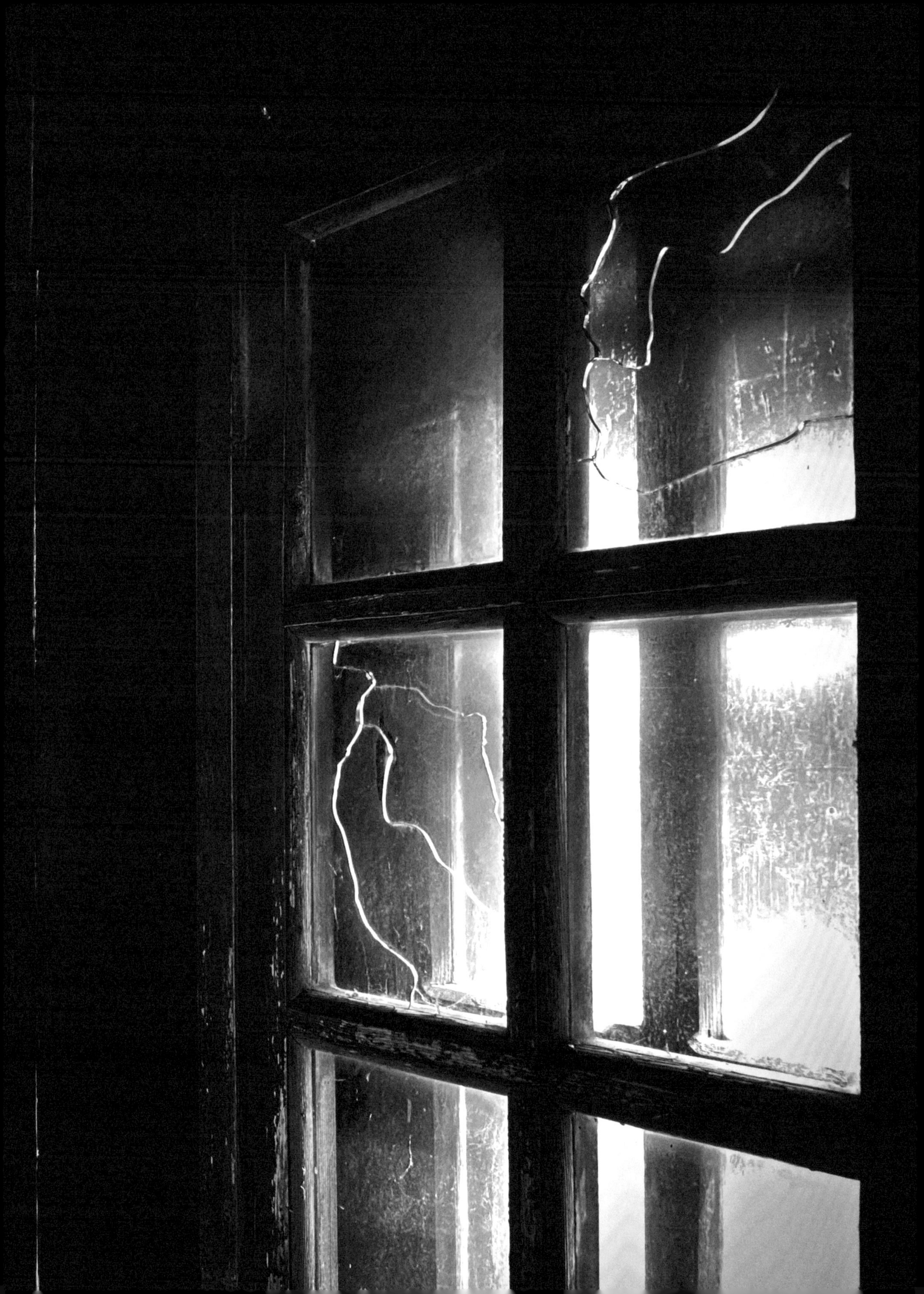

β'

23 July 2018
Weather Conditions

The forecasts issued to the Fire Department on the 22 July 2018, were *confirmed* by the Certificate of the Hellenic Meteorological Service with number 3219/1-8-18.

The certificate states that on 23 July 2018 and time 14:00 to 23:00, the weather was as follows:

Temporarily increased clouds in the afternoon hours.

Surface winds: West-Northwest, initially Fresh Breeze (5bf), strengthening after the first two hours of the above period to Strong Breeze (6 bf) to gusts of Near Gale (7bf) to Strong Gale (9bf) strength.

Temperature: From 31°C to 39°C.

Humidity: 15%–35%.

At the time of the fire at Ntaou Pentelis, the surface wind was the predicted West-Northwest, Strong Breeze (6bf), with Near Gale (7bf) to Strong Gale (9bf) gusts.

Information from the Report of Investigator of the Fire Department, Dimitrios Liotsios.

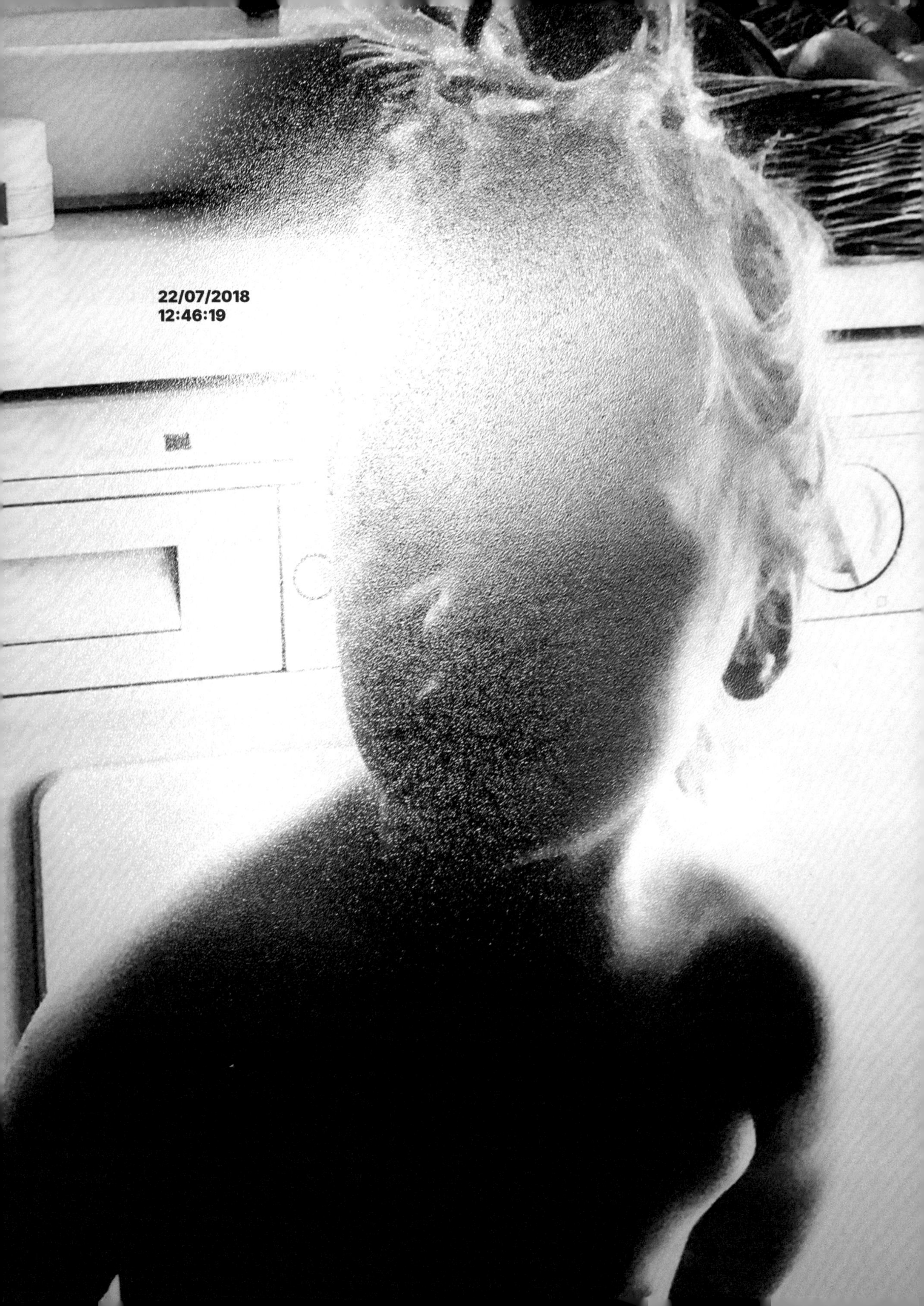
22/07/2018
12:46:19

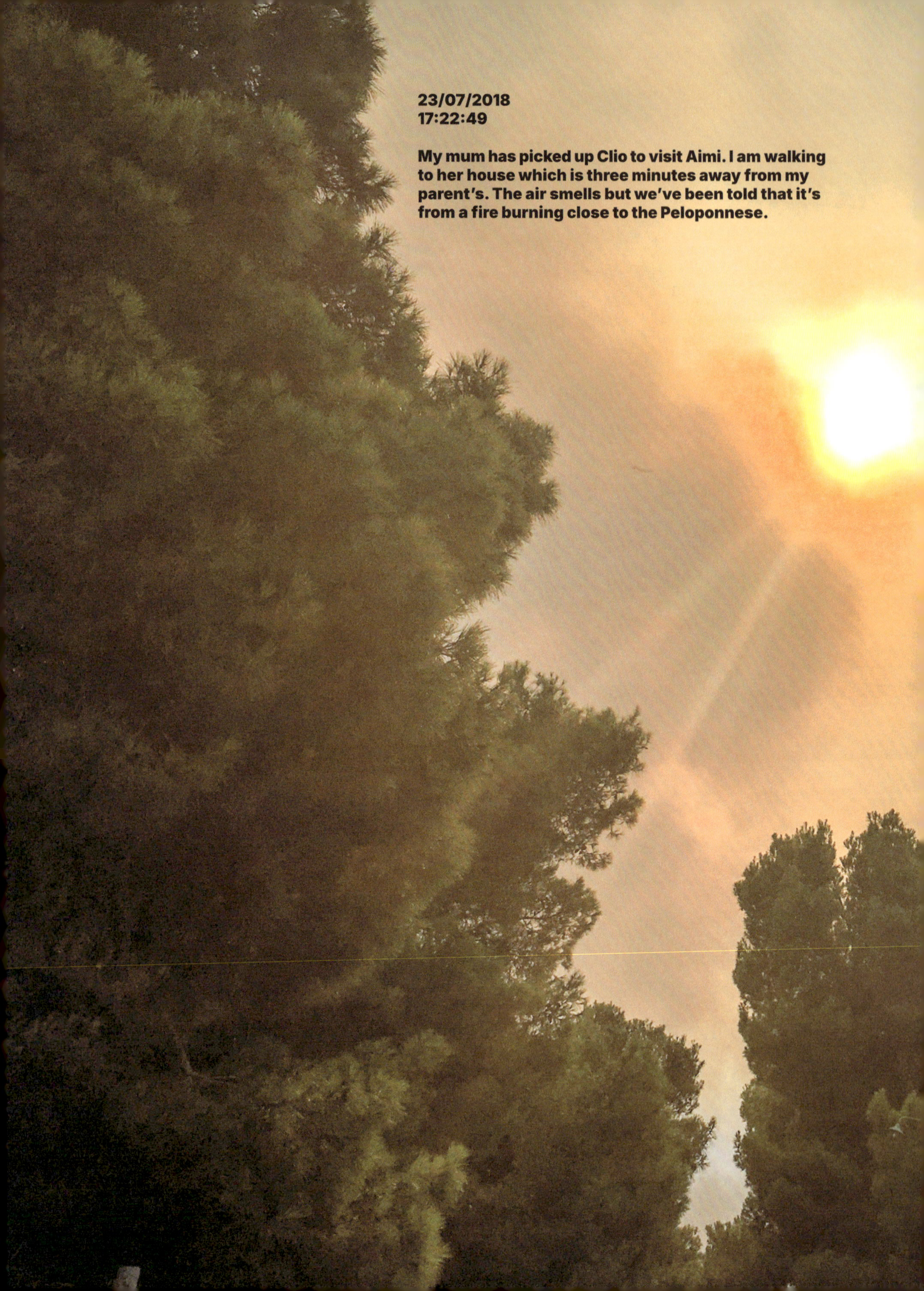

23/07/2018
17:22:49

My mum has picked up Clio to visit Aimi. I am walking to her house which is three minutes away from my parent's. The air smells but we've been told that it's from a fire burning close to the Peloponnese.

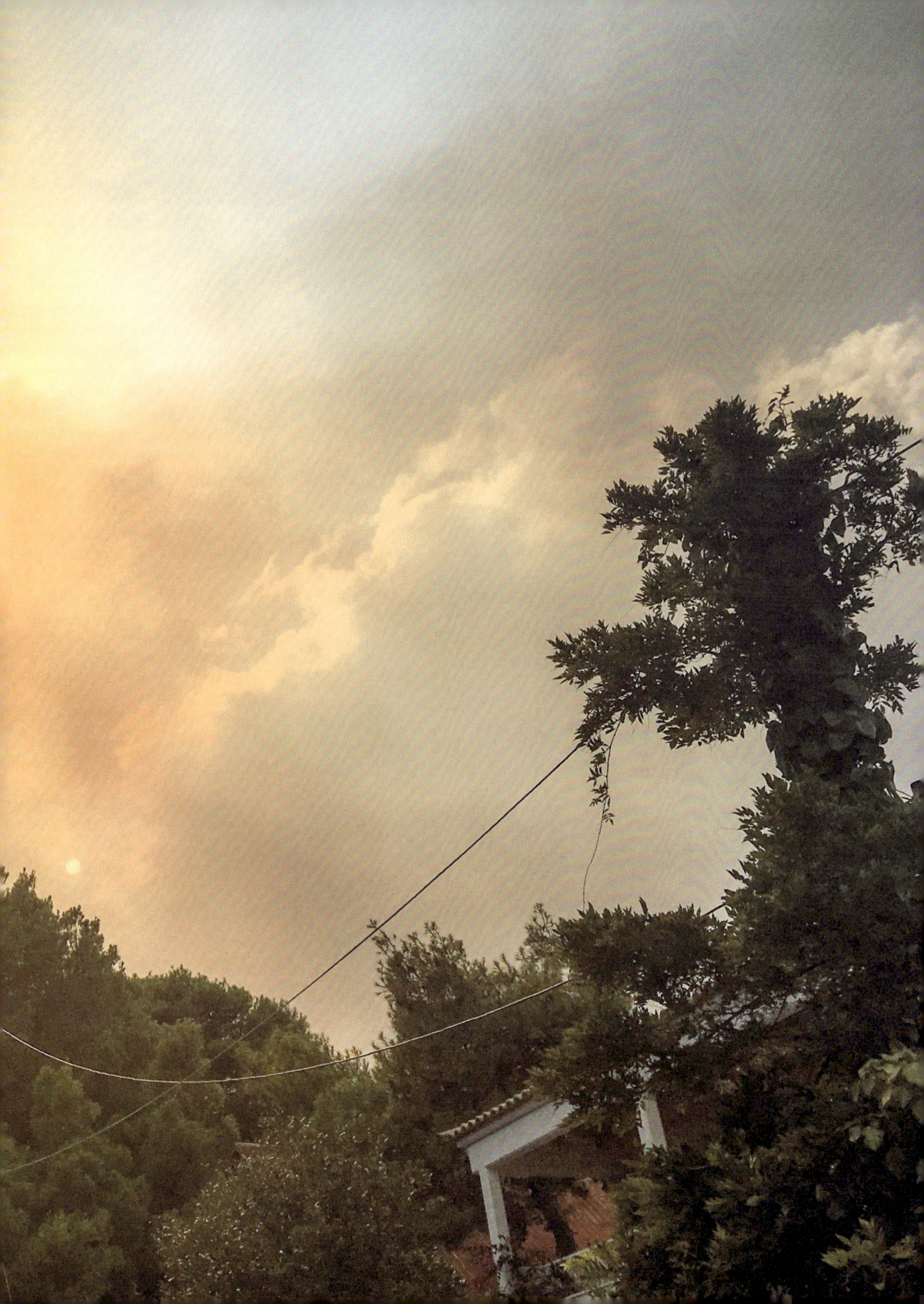

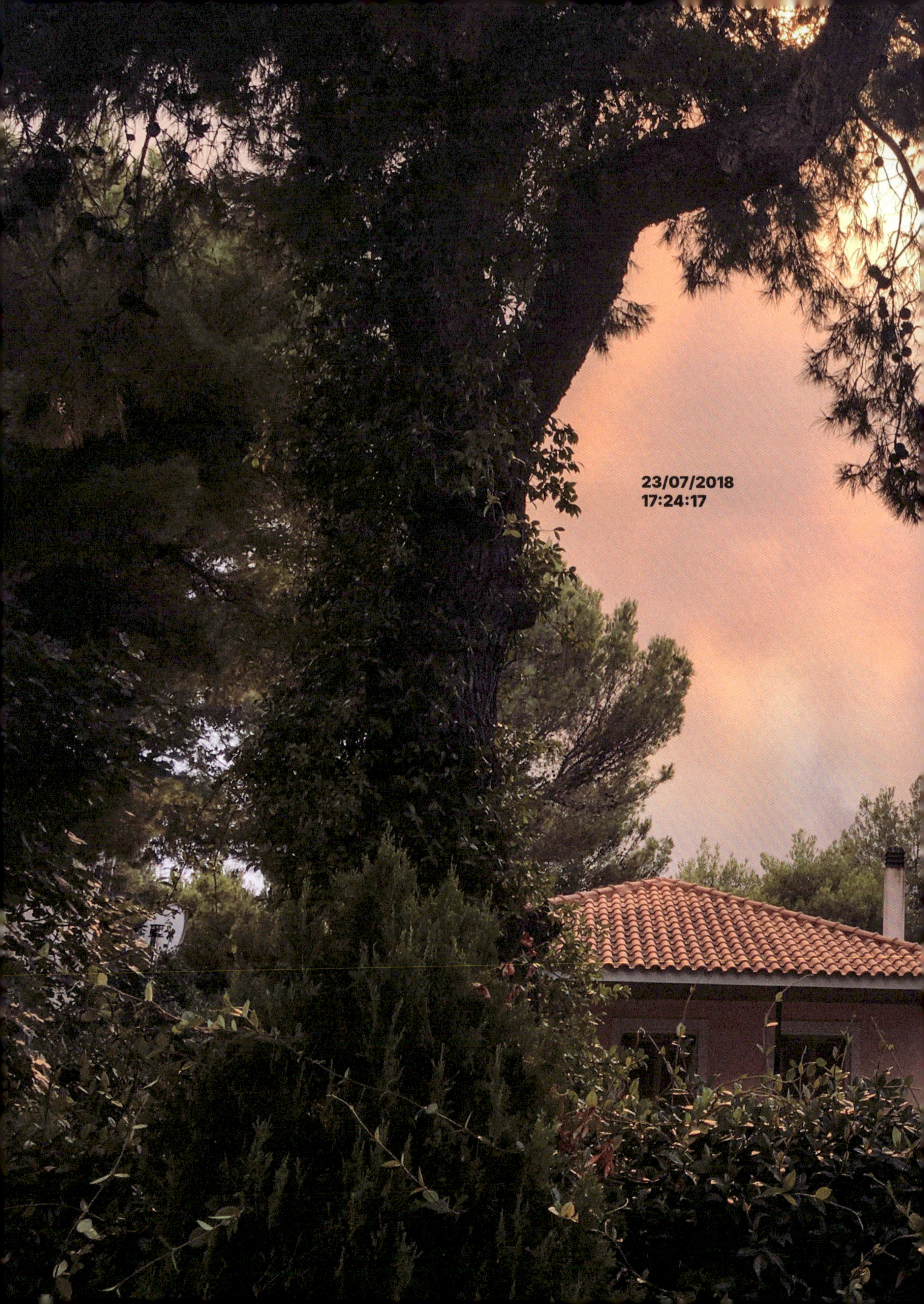
23/07/2018
17:24:17

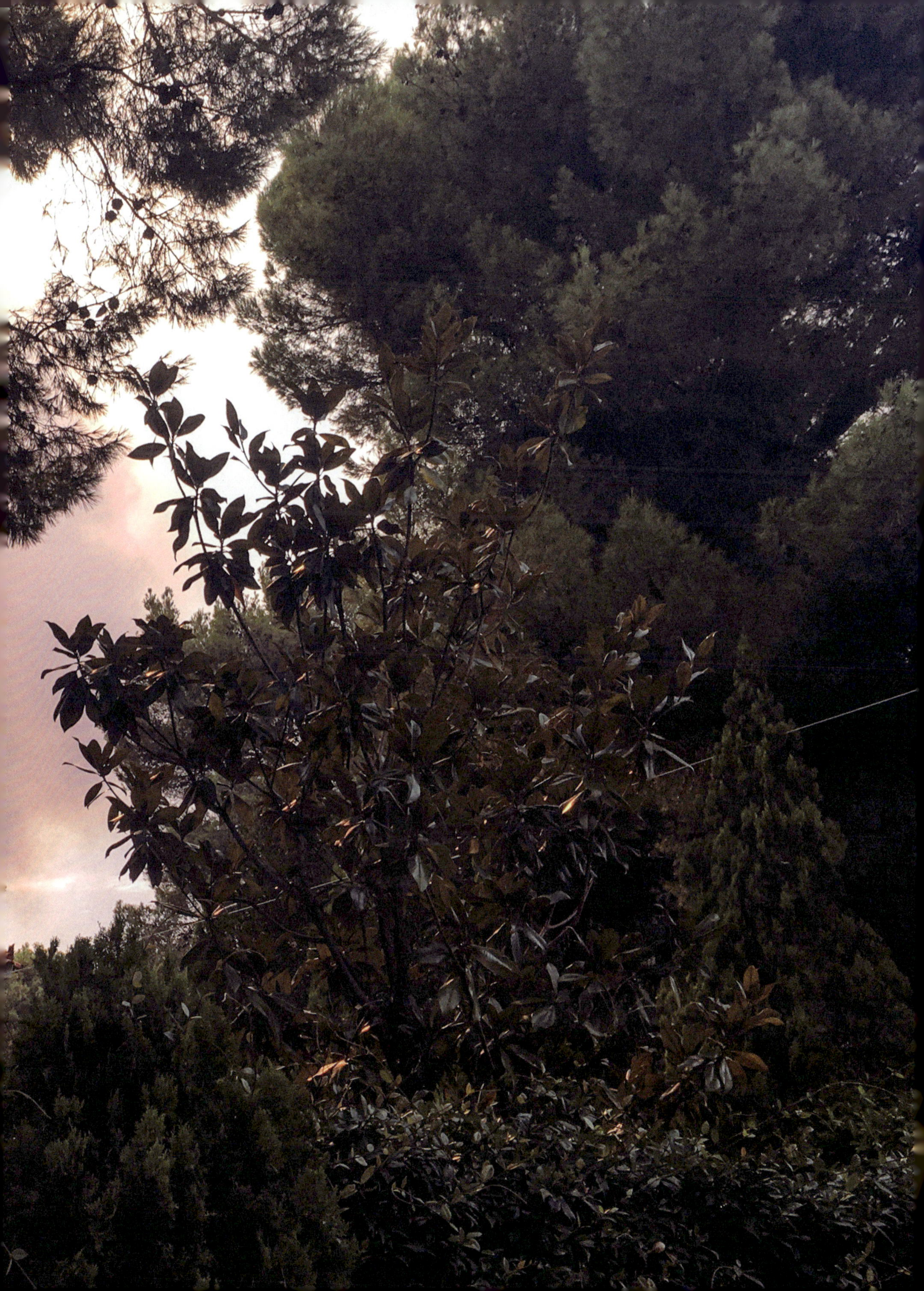

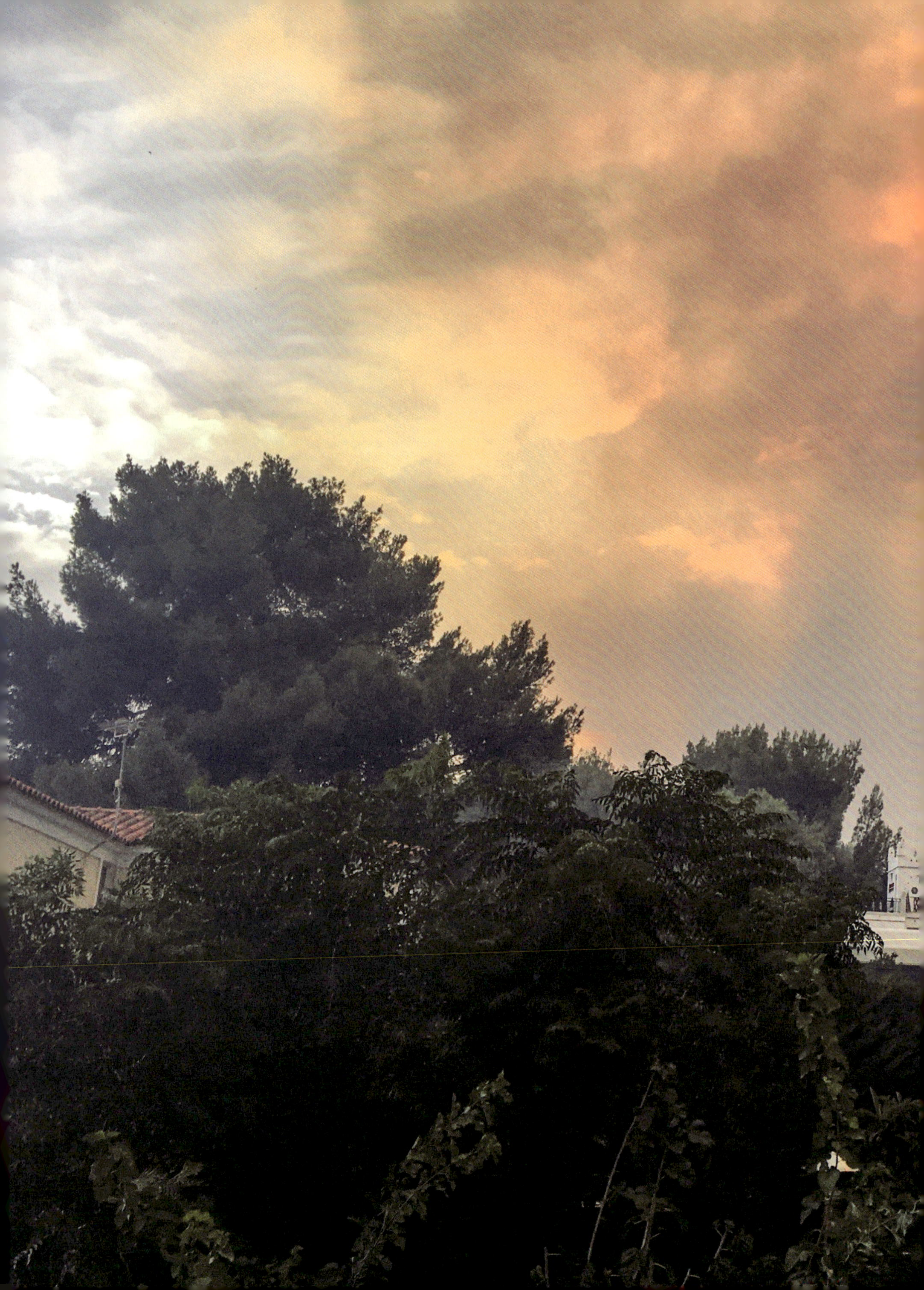

23/07/2018
18:01:00

The air and smoke are getting more intense. We hear nothing. No sirens, no bells, no cars, no people...In the news, nothing.

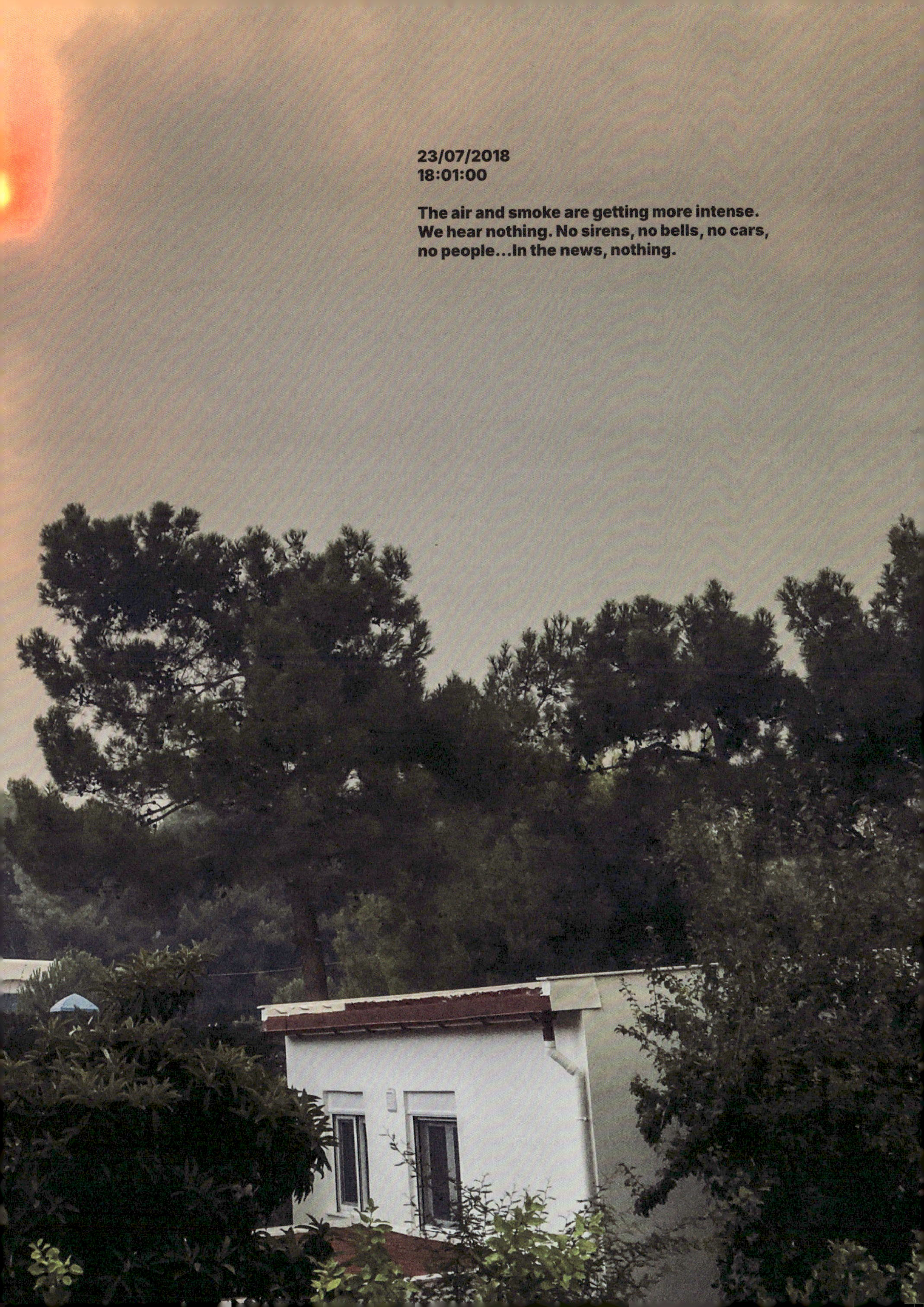

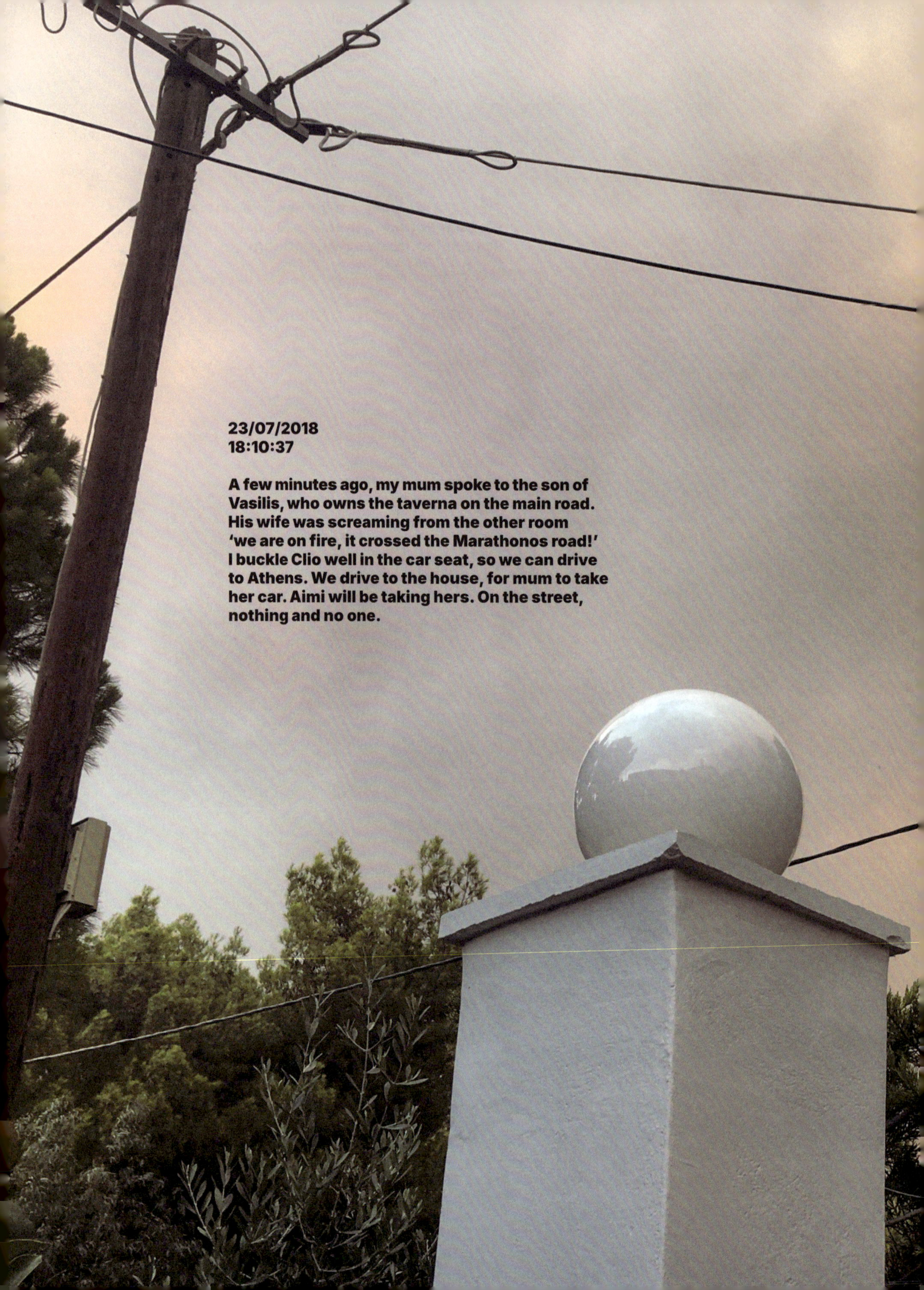

23/07/2018
18:10:37

A few minutes ago, my mum spoke to the son of Vasilis, who owns the taverna on the main road. His wife was screaming from the other room 'we are on fire, it crossed the Marathonos road!' I buckle Clio well in the car seat, so we can drive to Athens. We drive to the house, for mum to take her car. Aimi will be taking hers. On the street, nothing and no one.

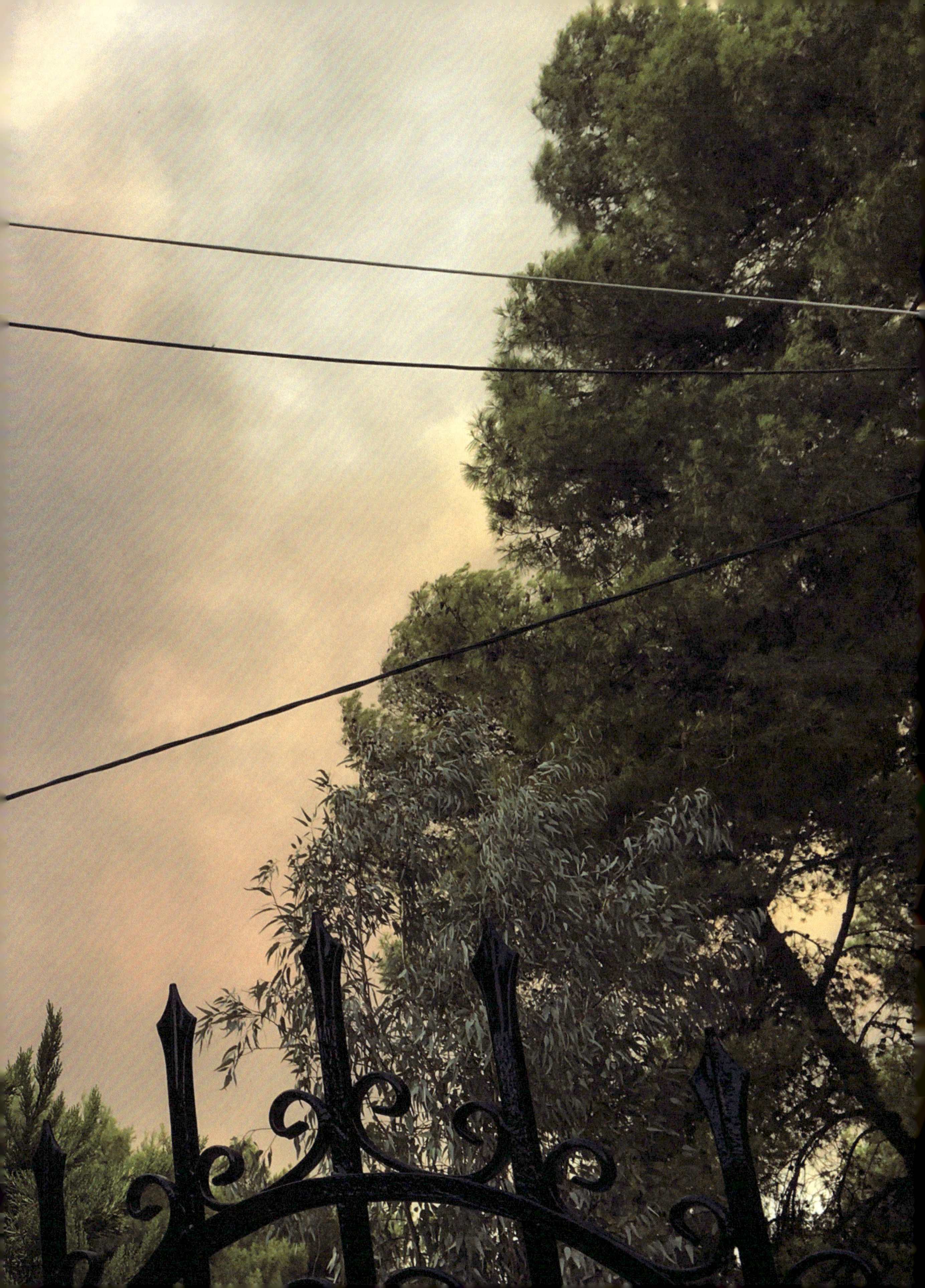

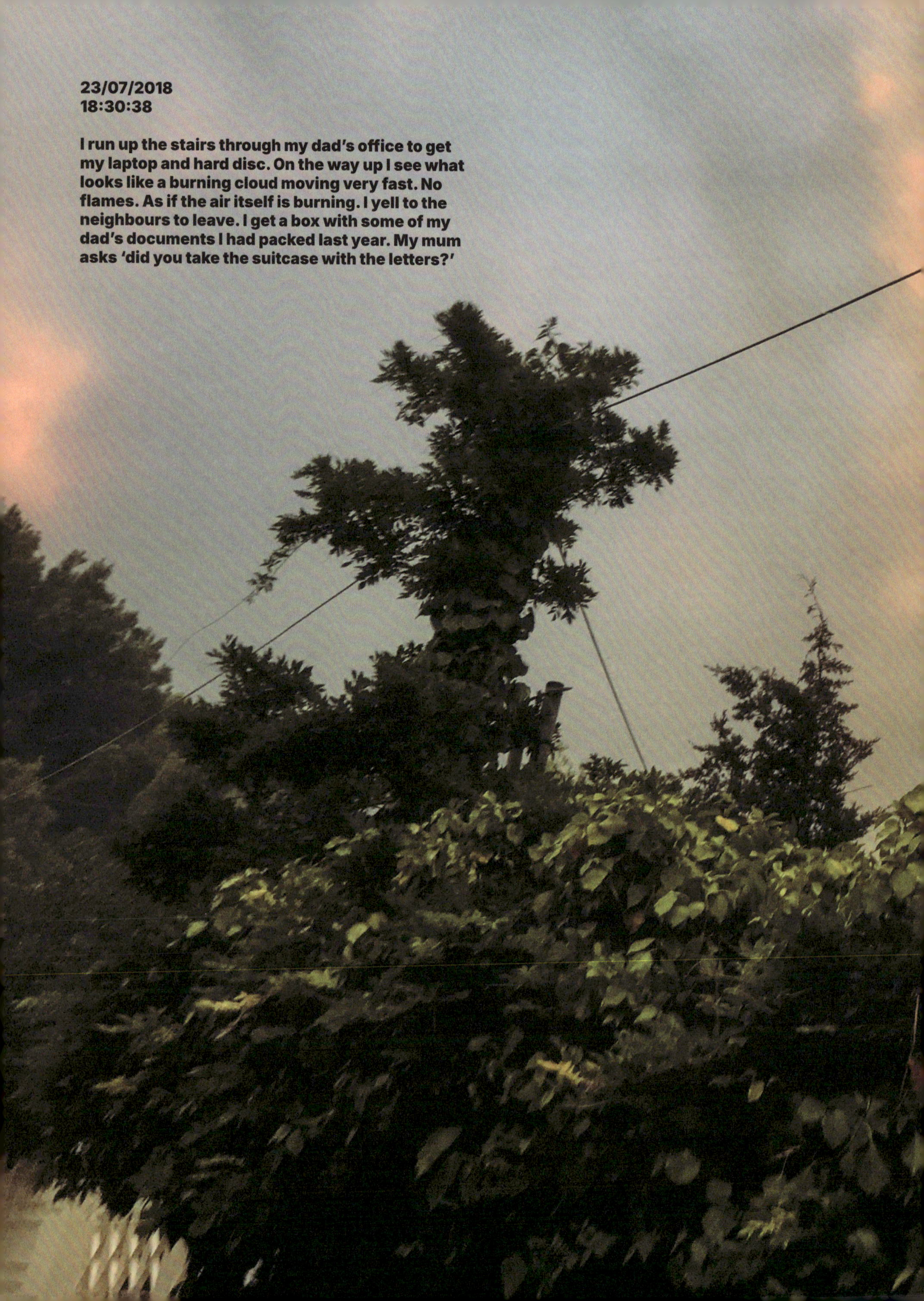

23/07/2018
18:30:38

I run up the stairs through my dad's office to get my laptop and hard disc. On the way up I see what looks like a burning cloud moving very fast. No flames. As if the air itself is burning. I yell to the neighbours to leave. I get a box with some of my dad's documents I had packed last year. My mum asks 'did you take the suitcase with the letters?'

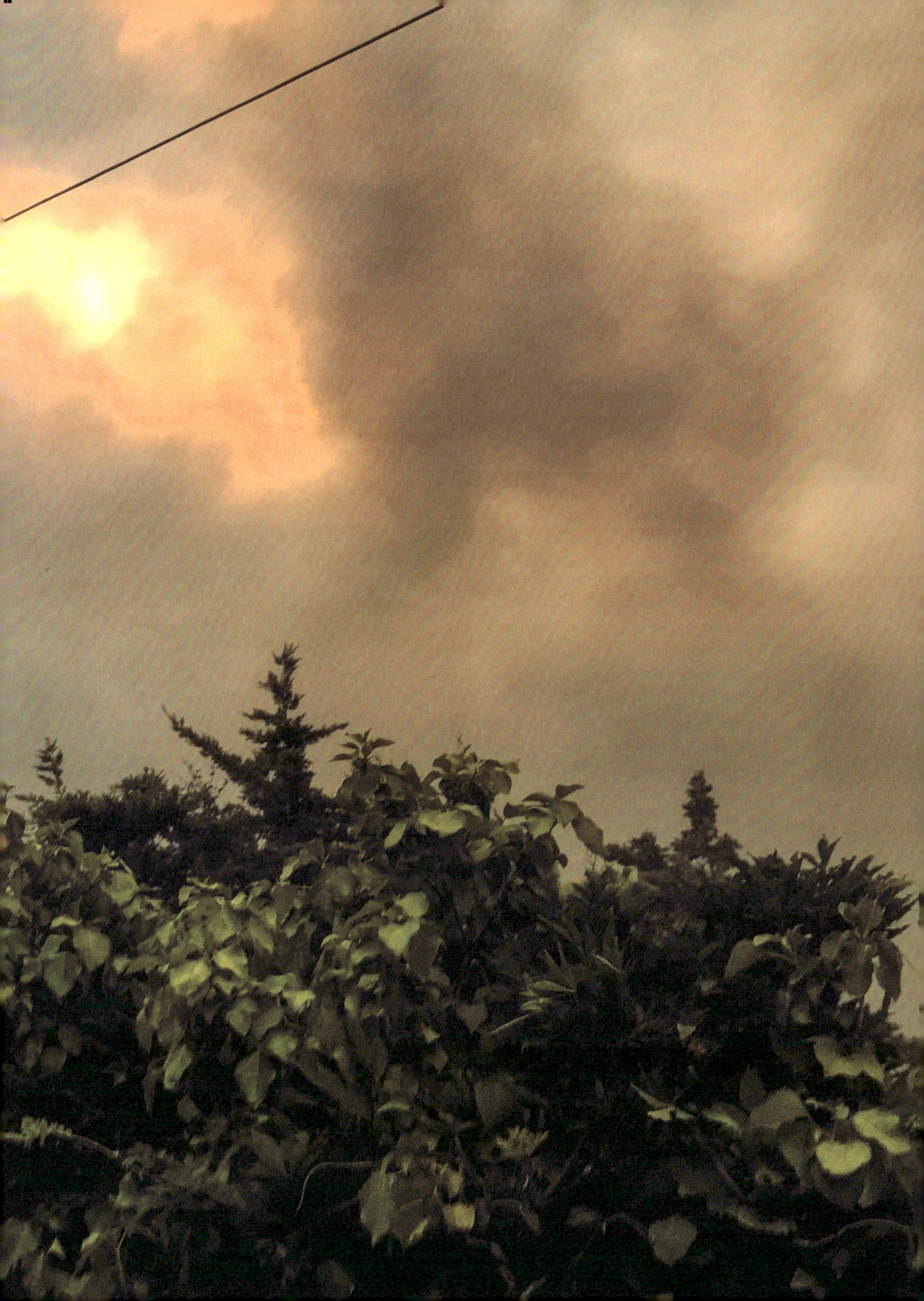

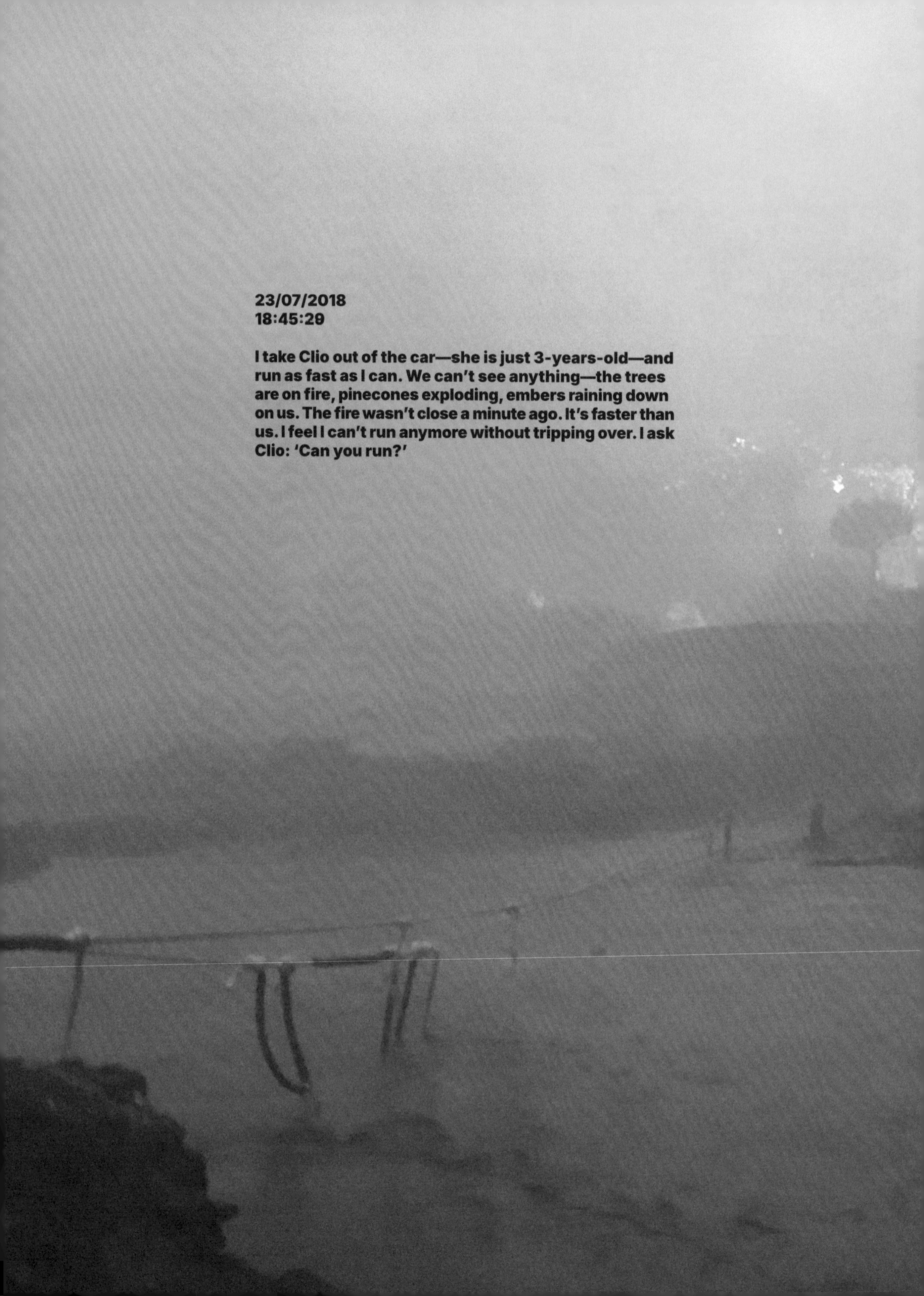

23/07/2018
18:45:29

I take Clio out of the car—she is just 3-years-old—and run as fast as I can. We can't see anything—the trees are on fire, pinecones exploding, embers raining down on us. The fire wasn't close a minute ago. It's faster than us. I feel I can't run anymore without tripping over. I ask Clio: 'Can you run?'

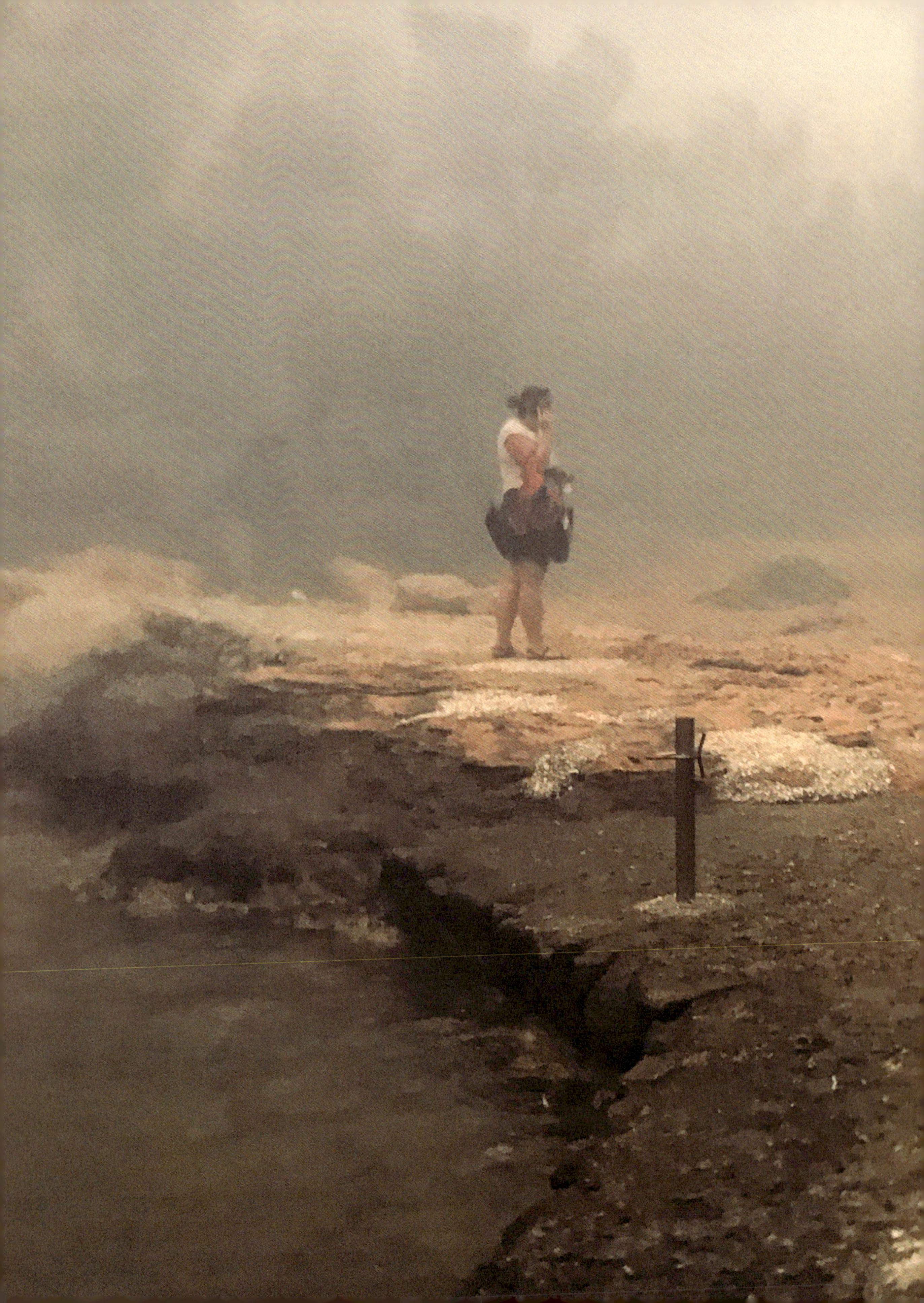

23/07/2018
18:45:35

We are out of breath. I did not take any water. A lady is holding two small bottles. I ask for a sip for Clio and I. She hands me one, and says I should keep it for the child. I never find her to thank her later.

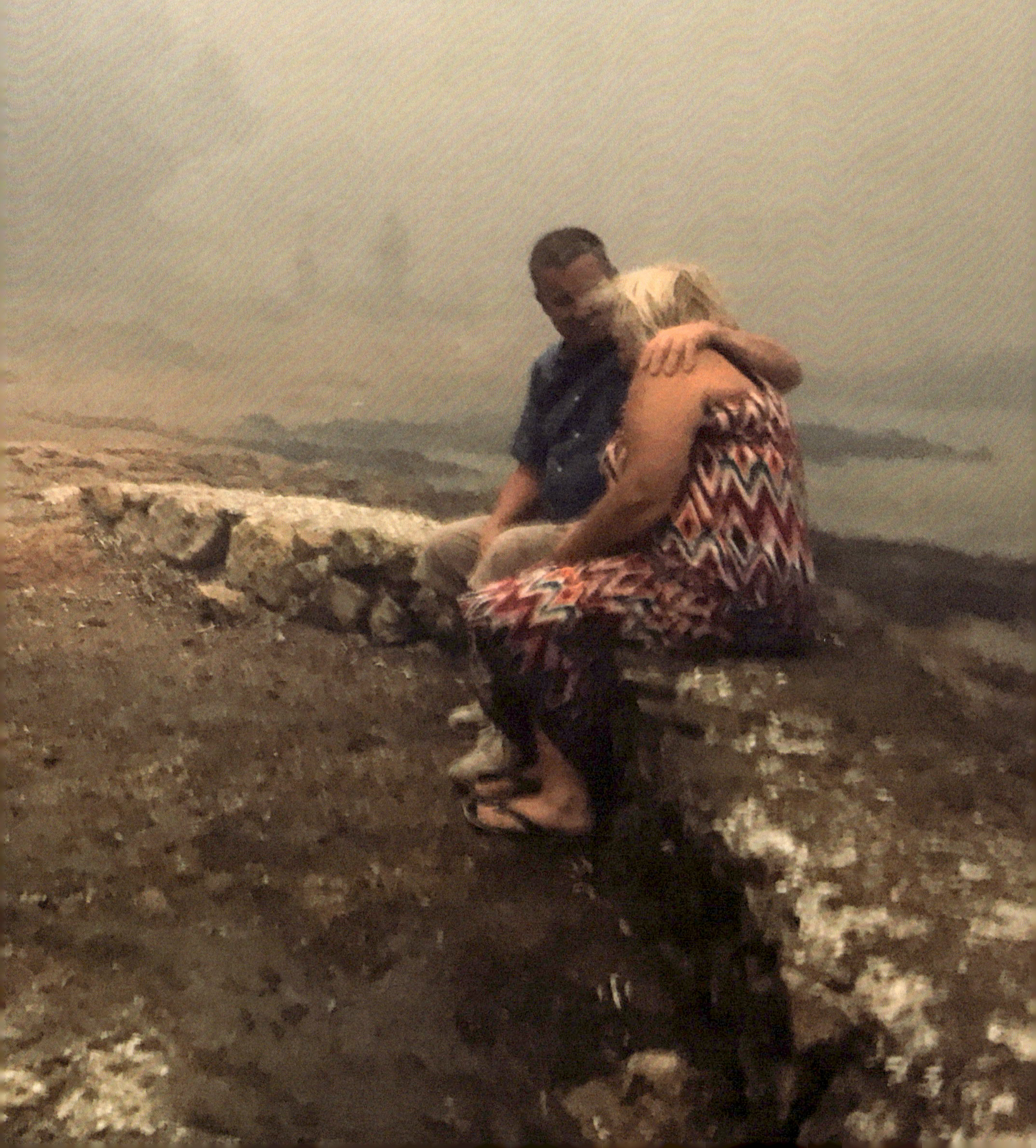

23/07/2018
18:45:35

My mum said she was right behind us. She was getting out of the car and coming? The fire was in front of her. Did I leave her behind? I left her behind. I need to go get her. Where will I leave Clio?

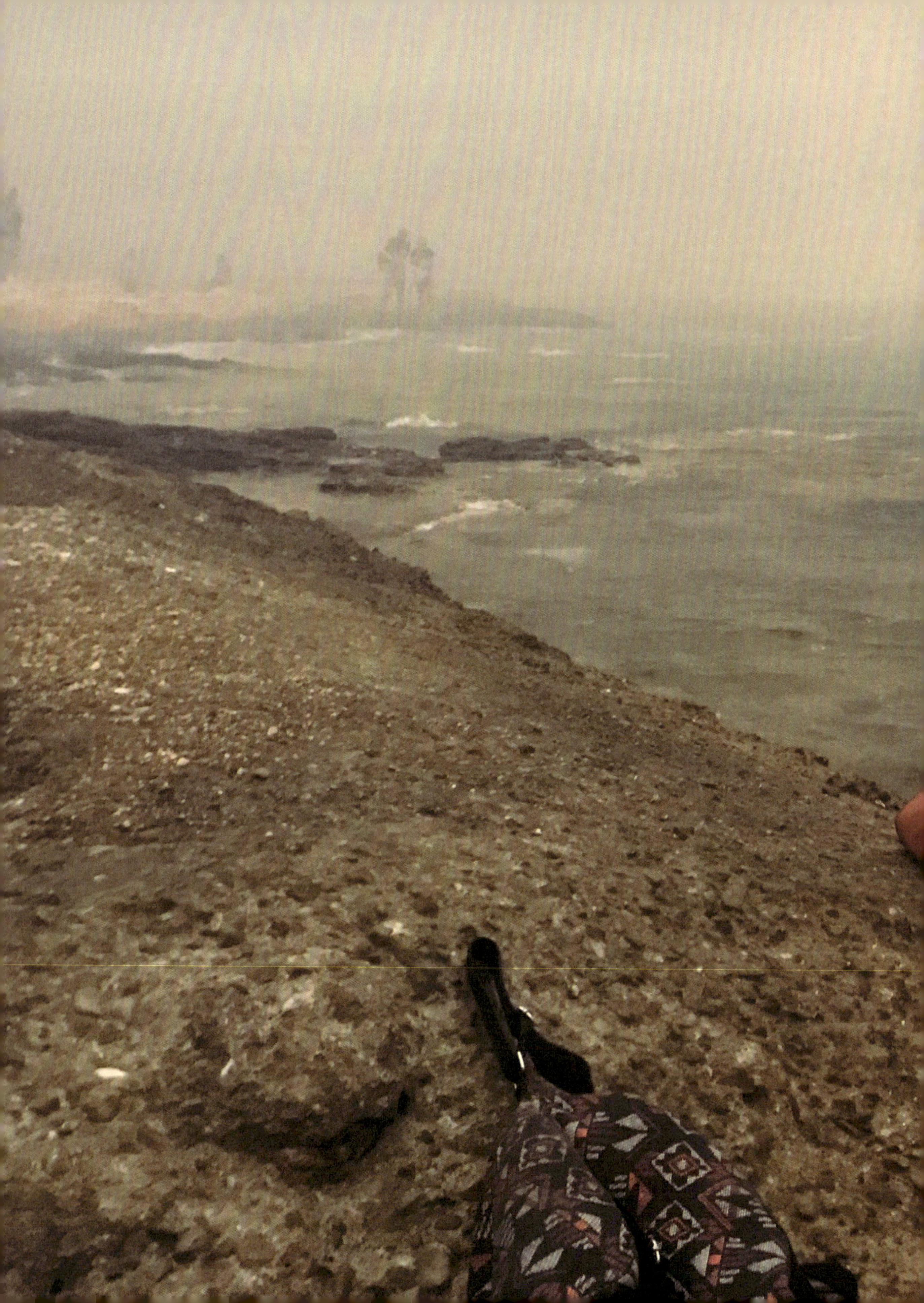

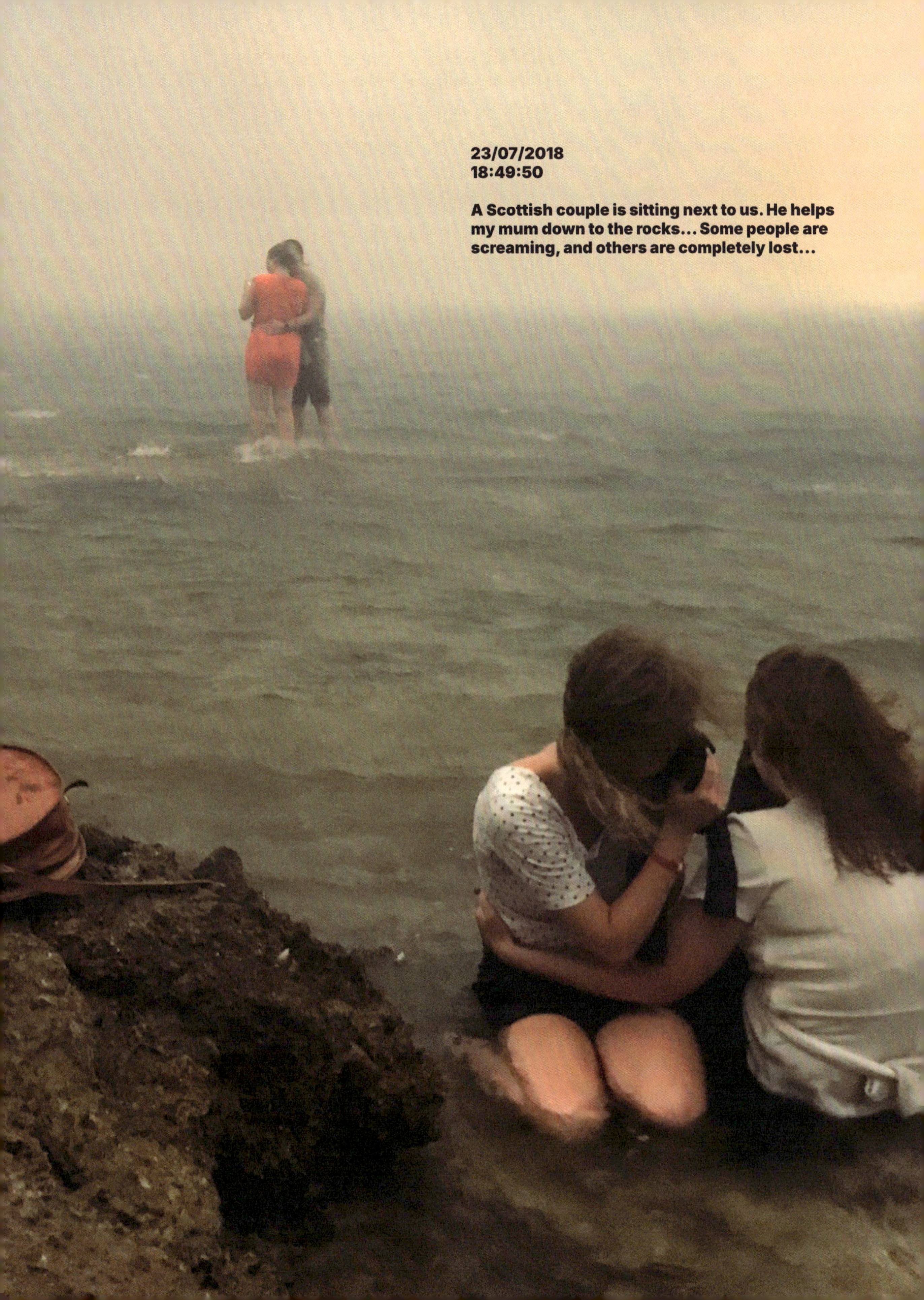

23/07/2018
18:49:50

A Scottish couple is sitting next to us. He helps my mum down to the rocks... Some people are screaming, and others are completely lost...

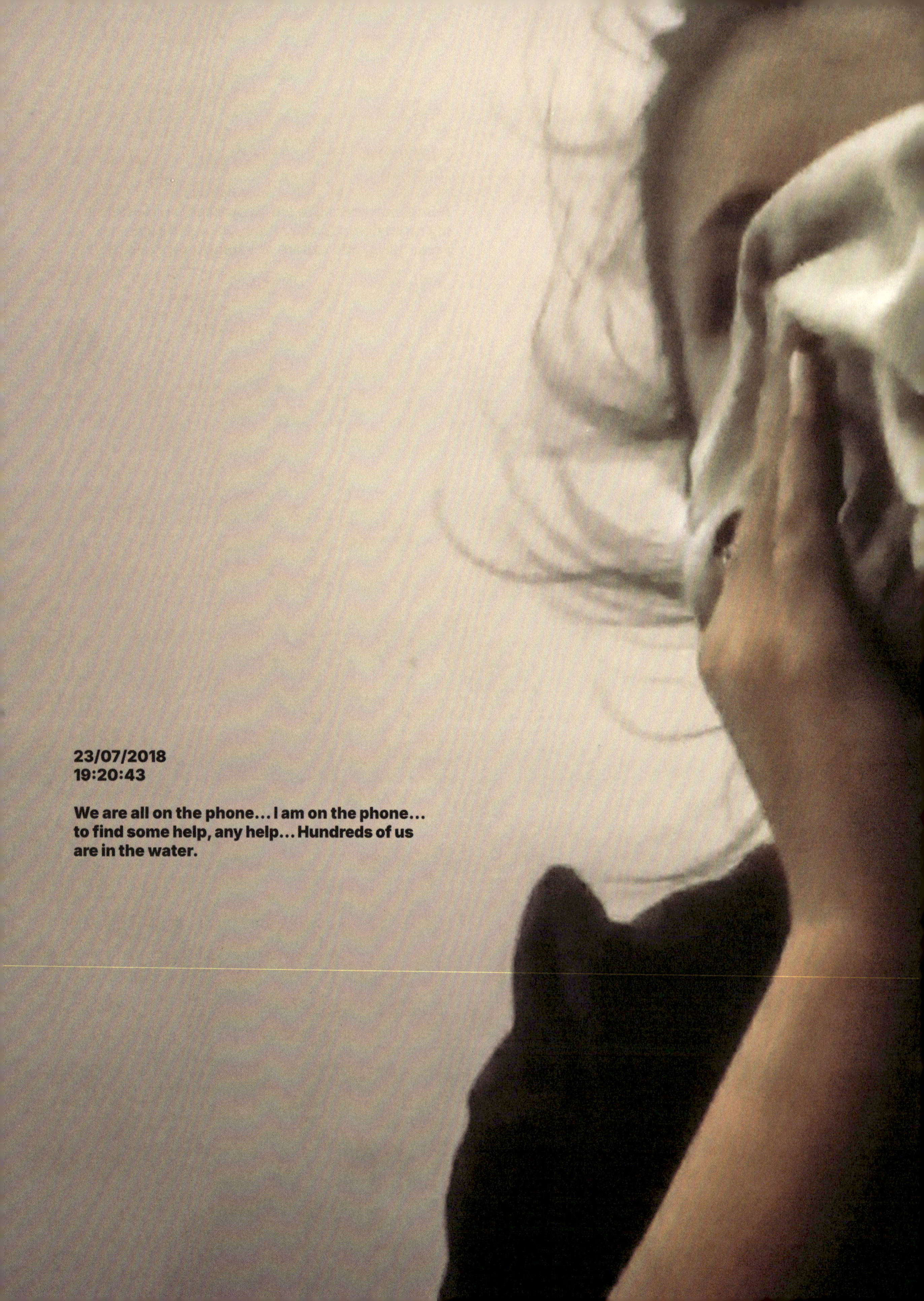

23/07/2018
19:20:43

We are all on the phone... I am on the phone... to find some help, any help... Hundreds of us are in the water.

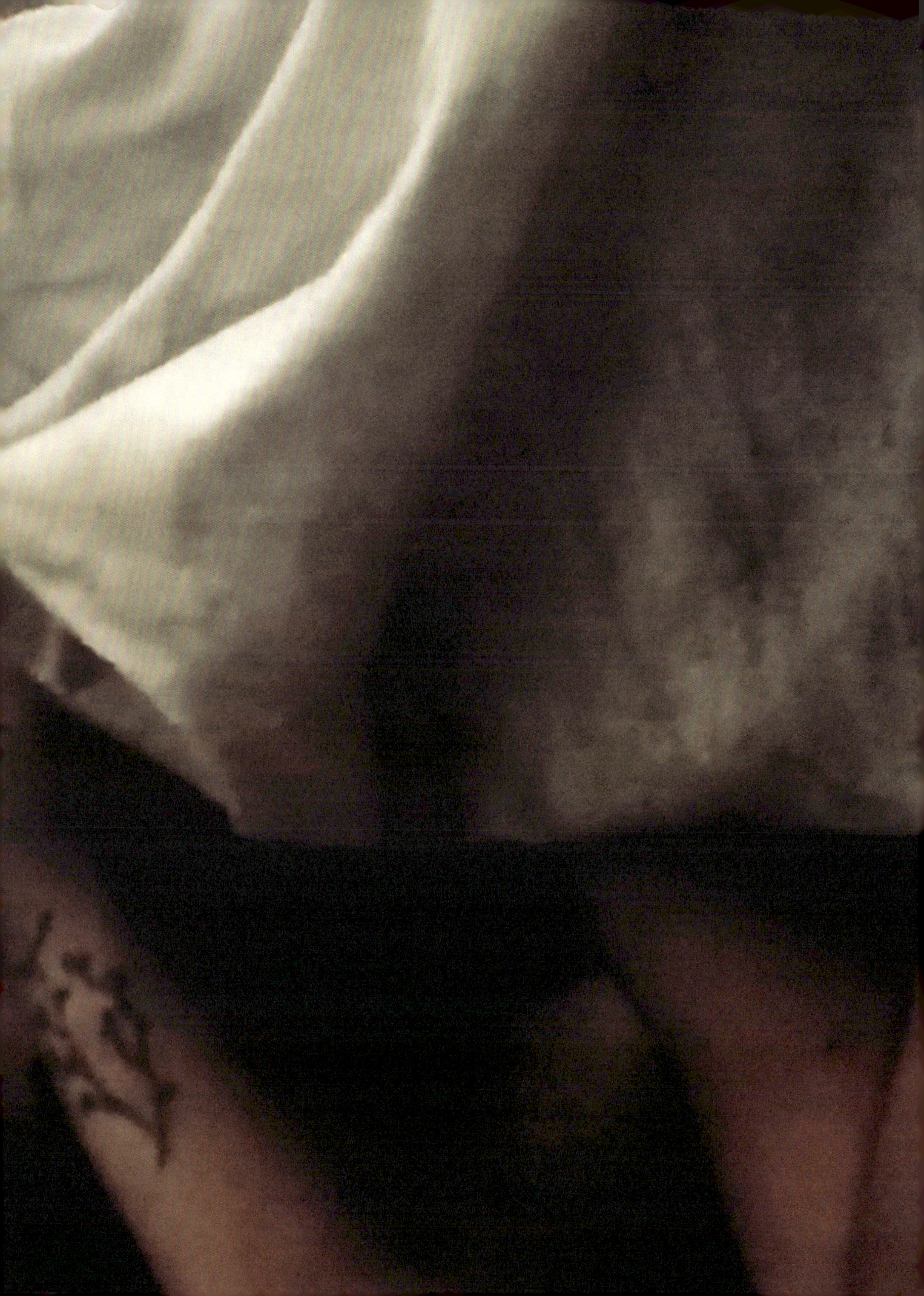

23/07/2018
19:20:45

We sing nursery songs with Clio to keep her distracted.

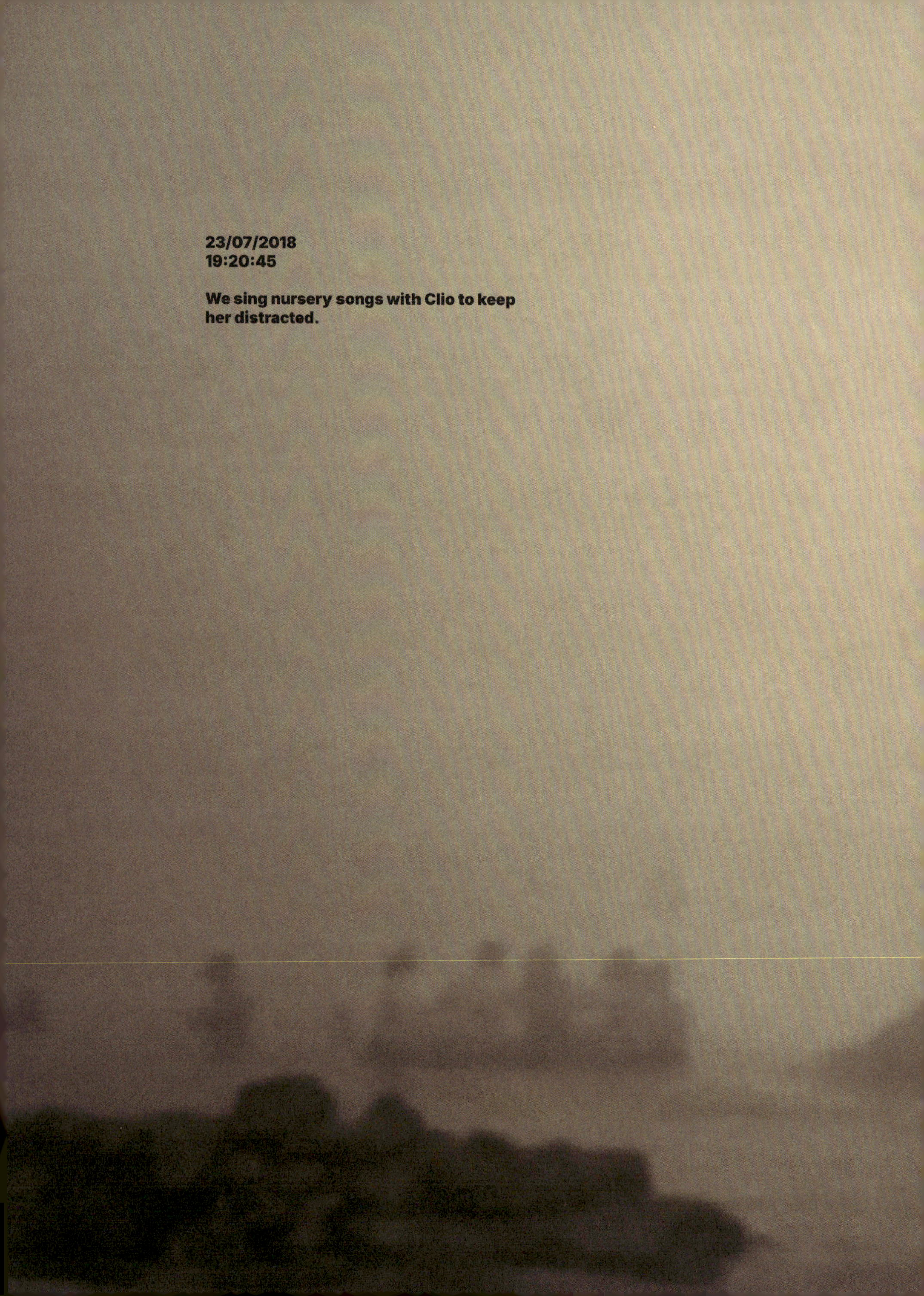

23/07/2018
19:56:26

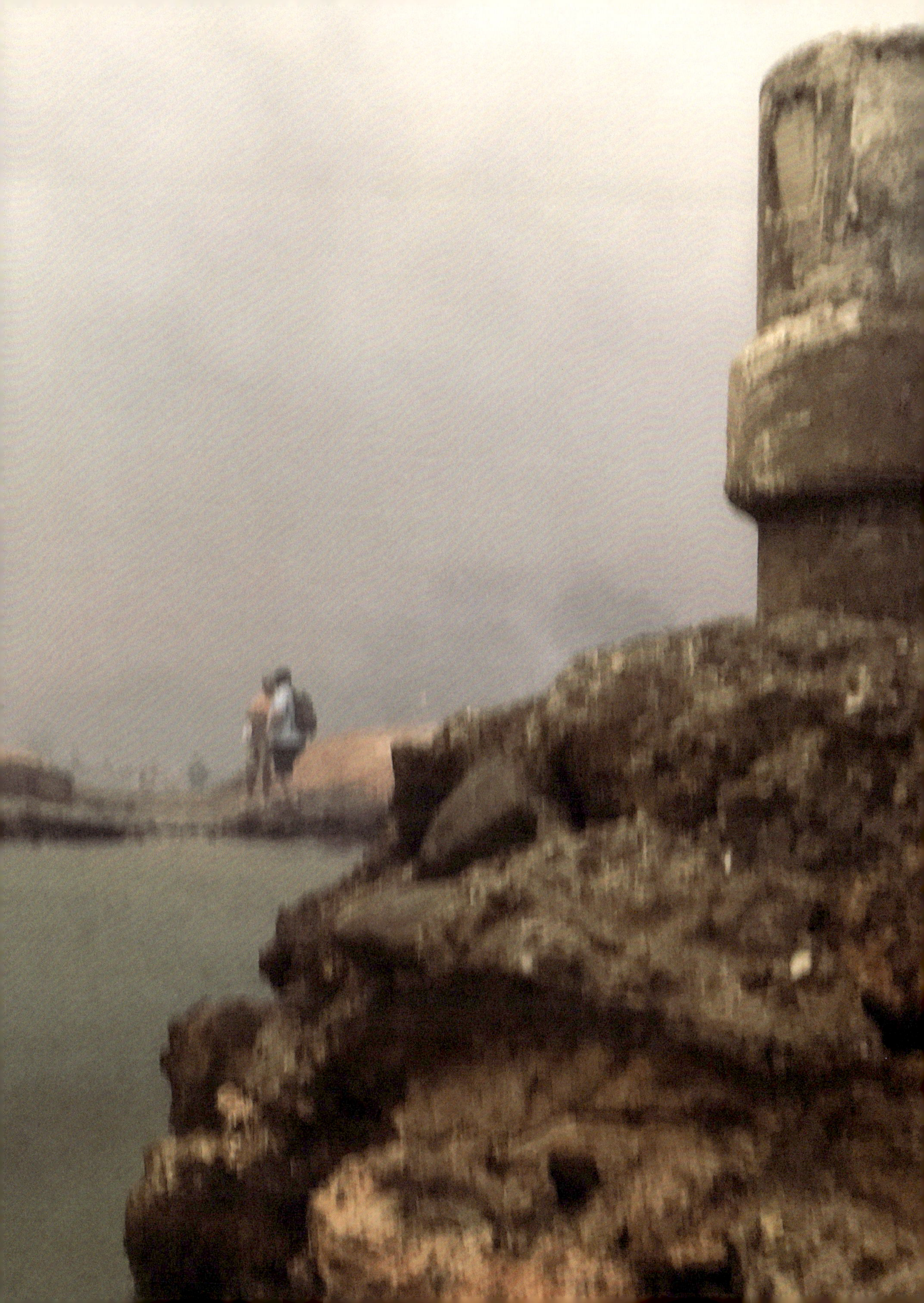

23/07/2018
19:56:38

I sit on this rock for three hours half submerged in the water. Clio sits on my lap. The rock has cut deep into my skin all the way down to my knees. I feel no pain.

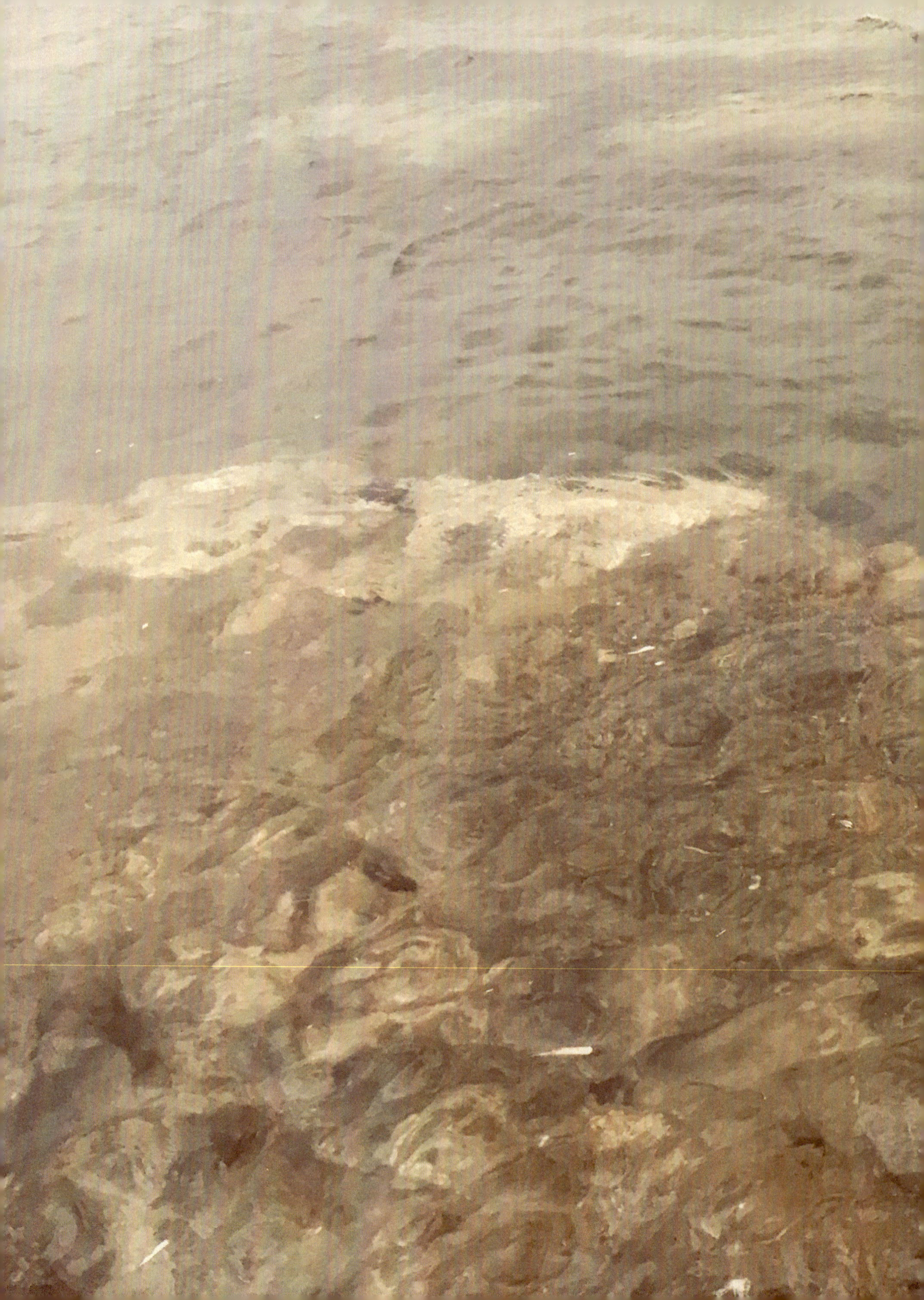

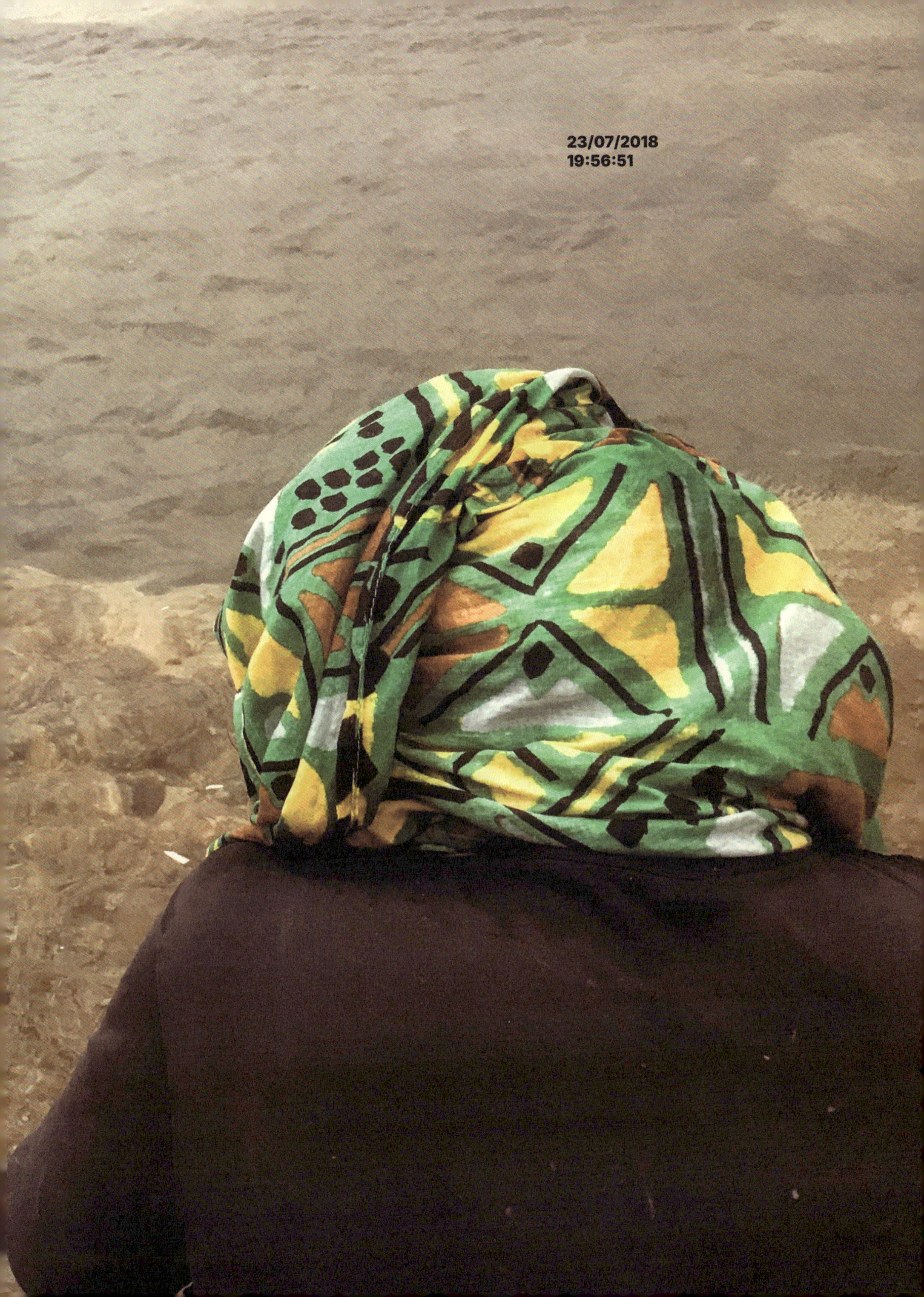
23/07/2018
19:56:51

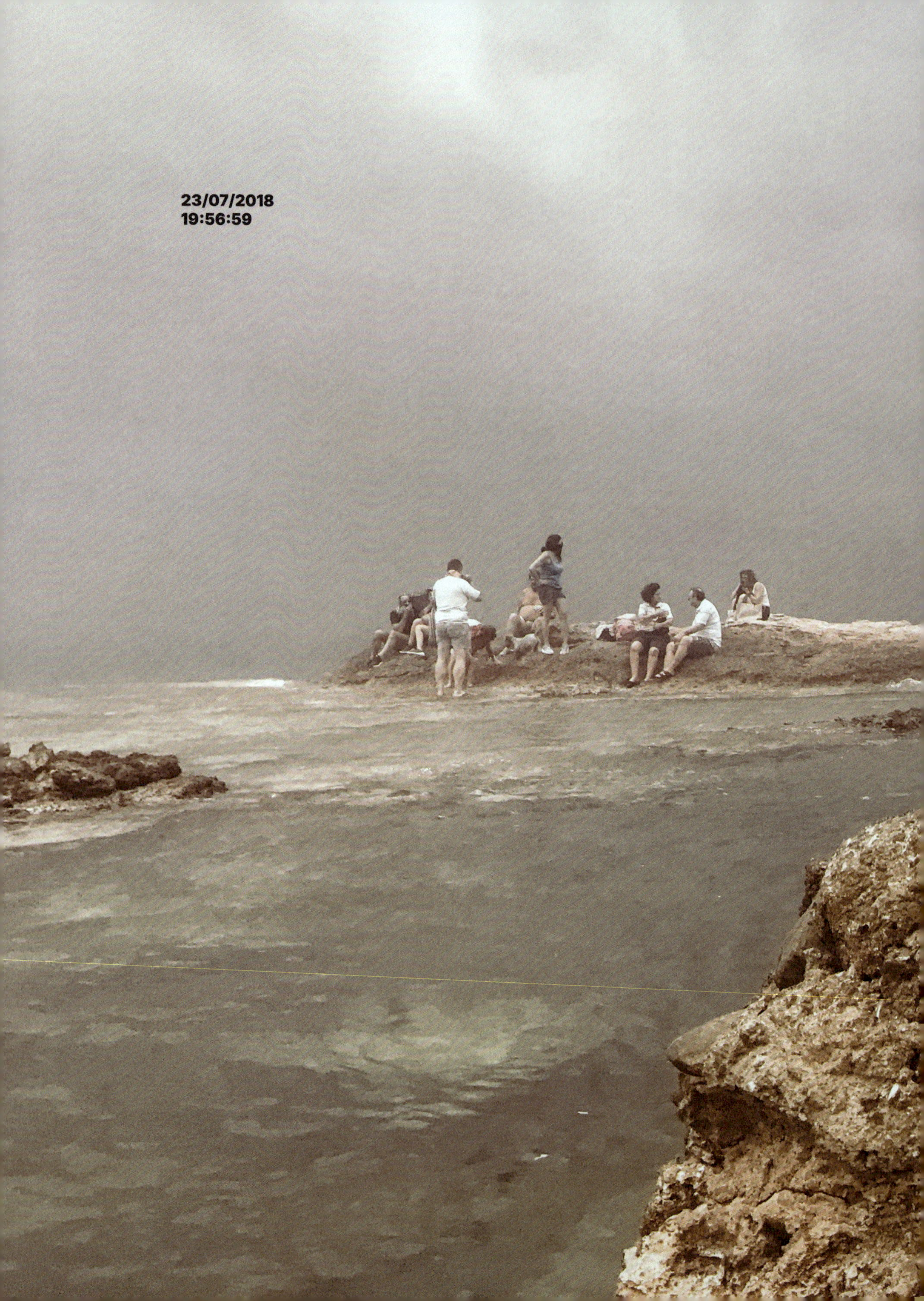
23/07/2018
19:56:59

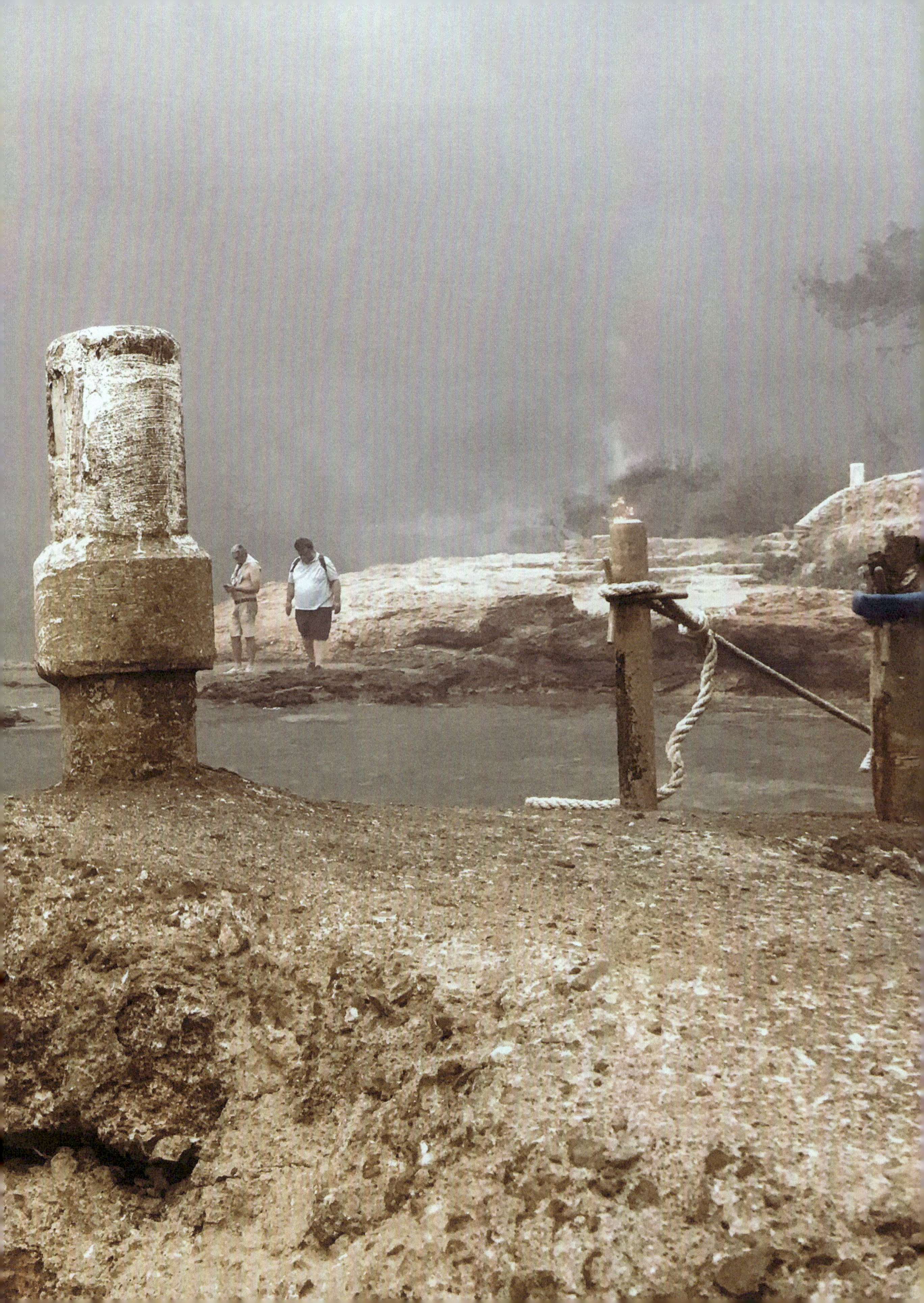

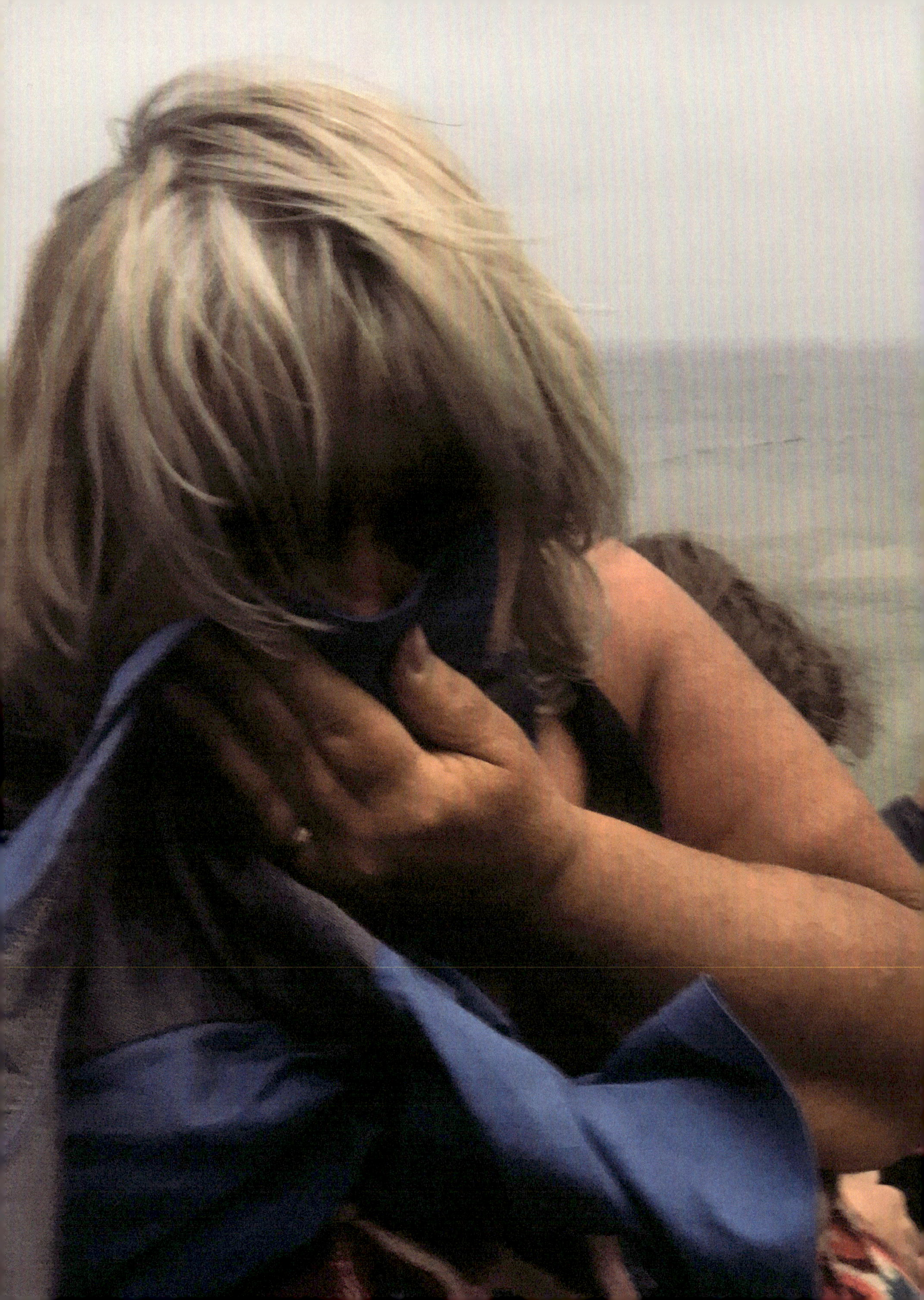

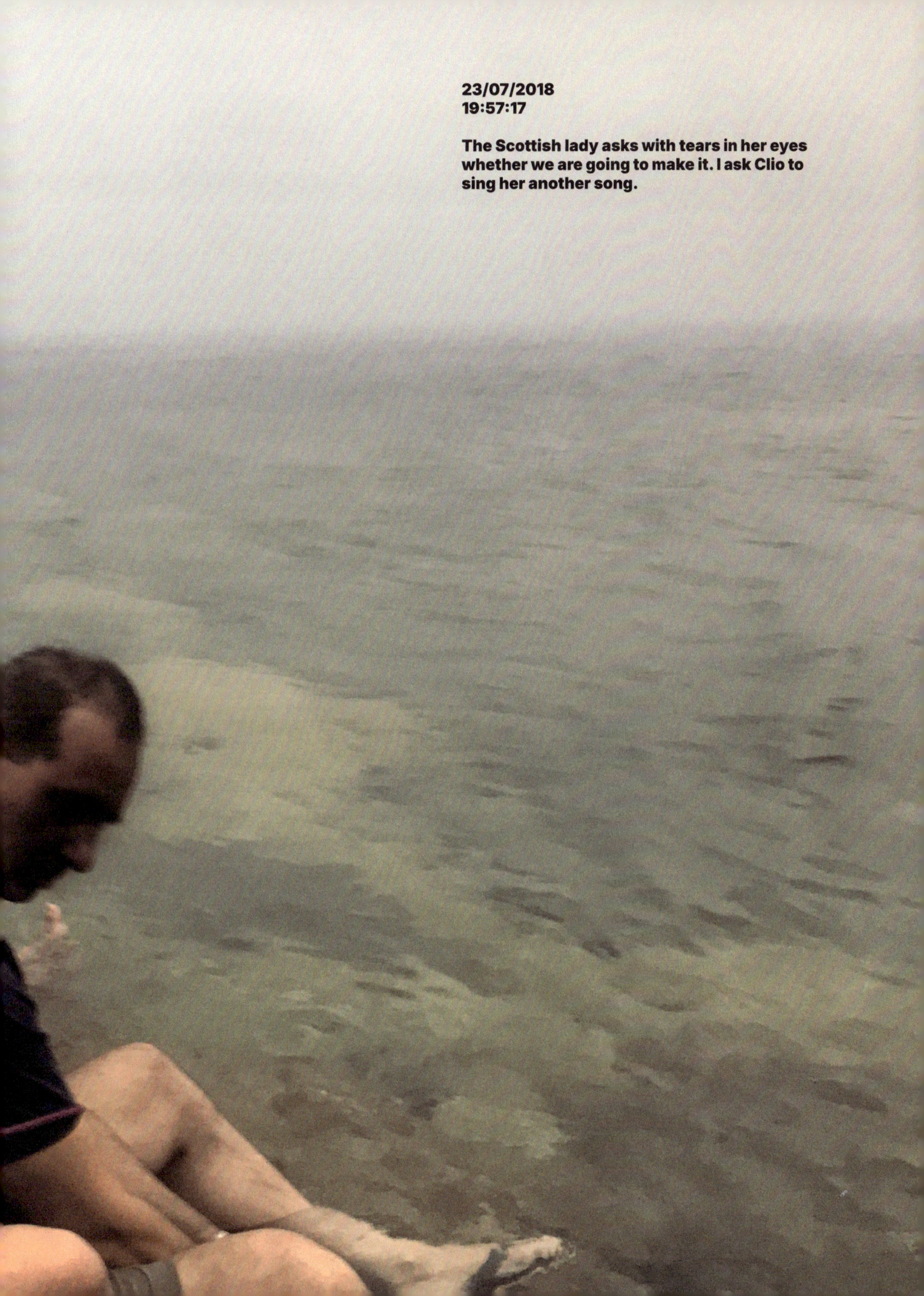

23/07/2018
19:57:17

The Scottish lady asks with tears in her eyes whether we are going to make it. I ask Clio to sing her another song.

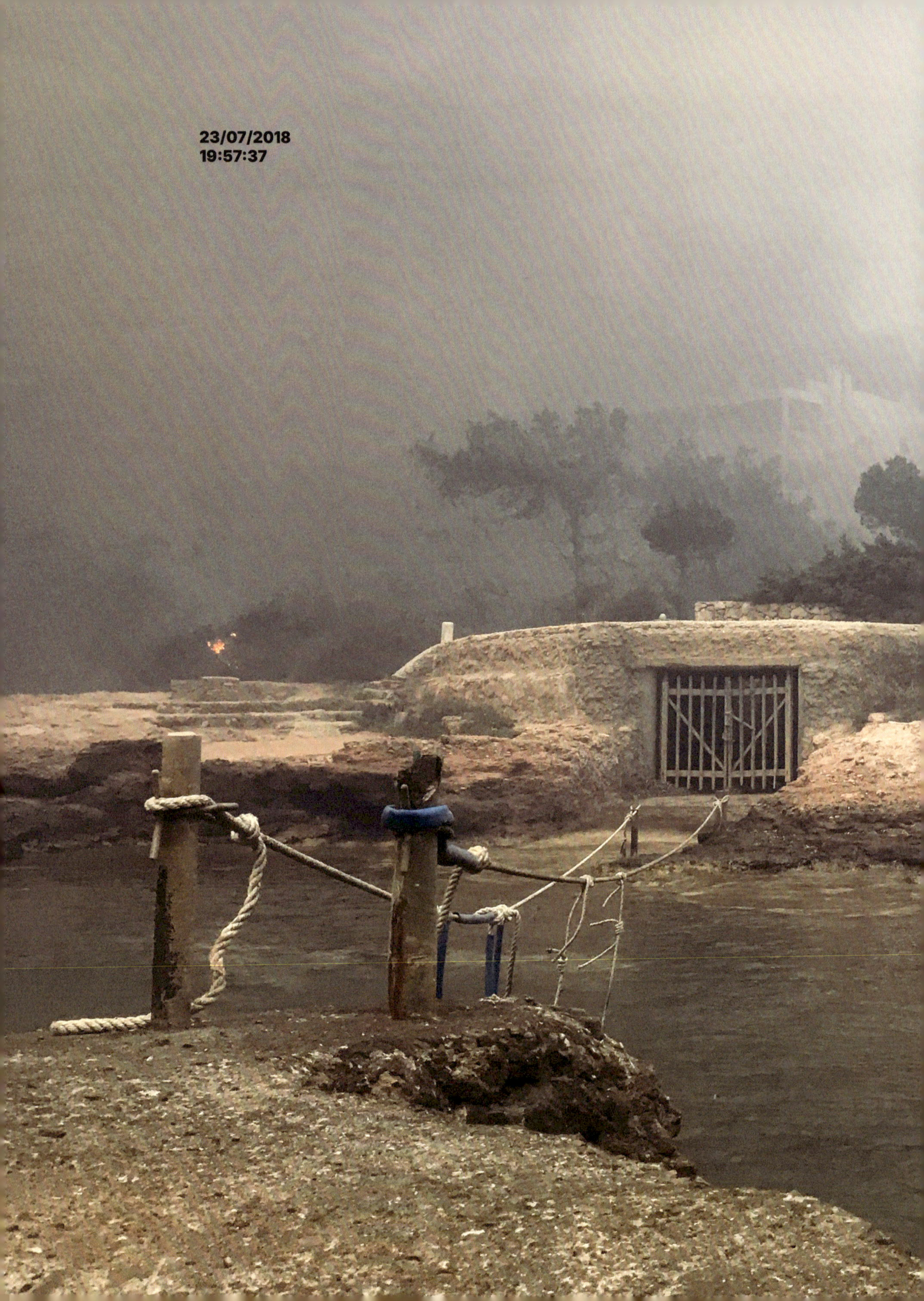
23/07/2018
19:57:37

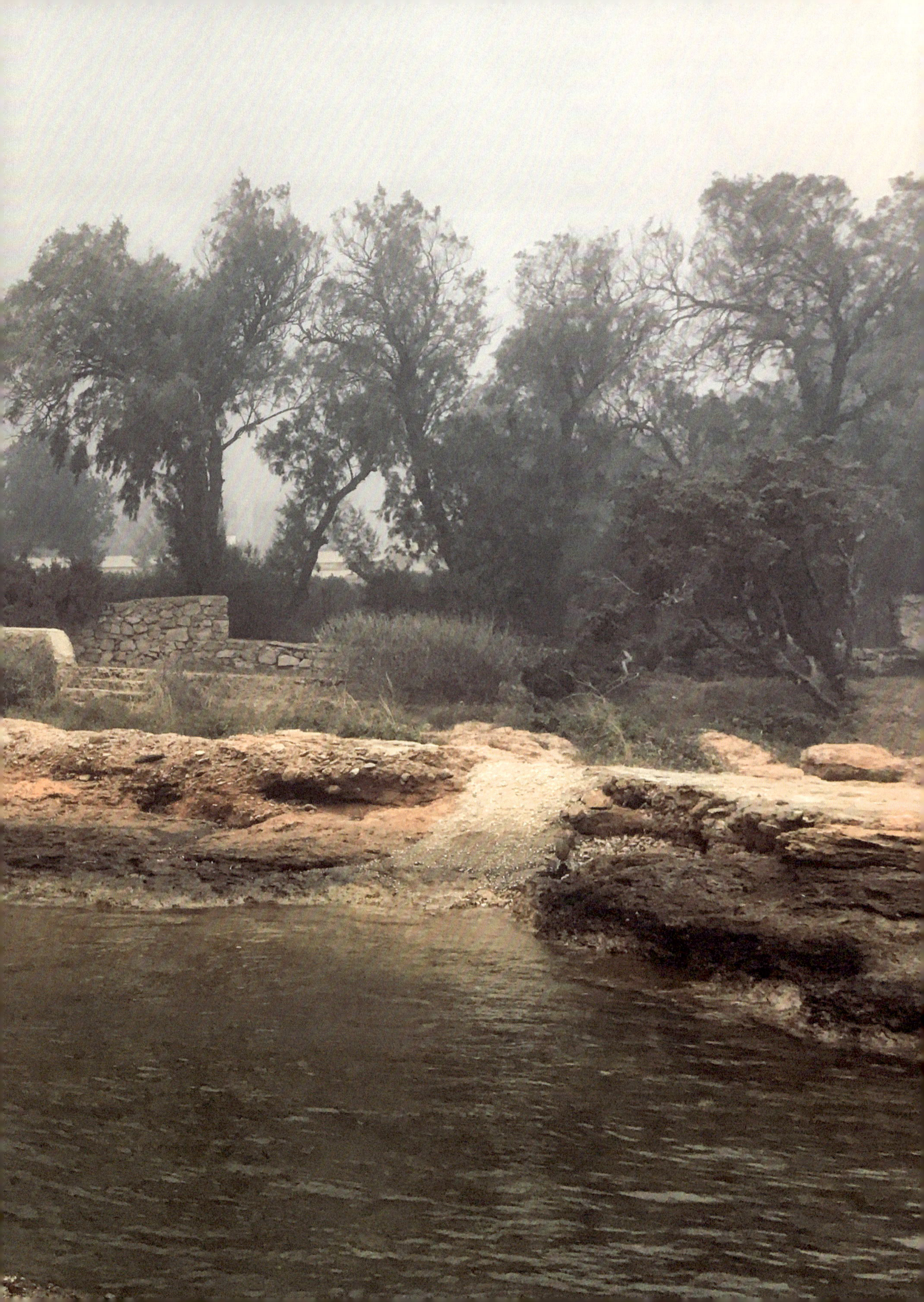

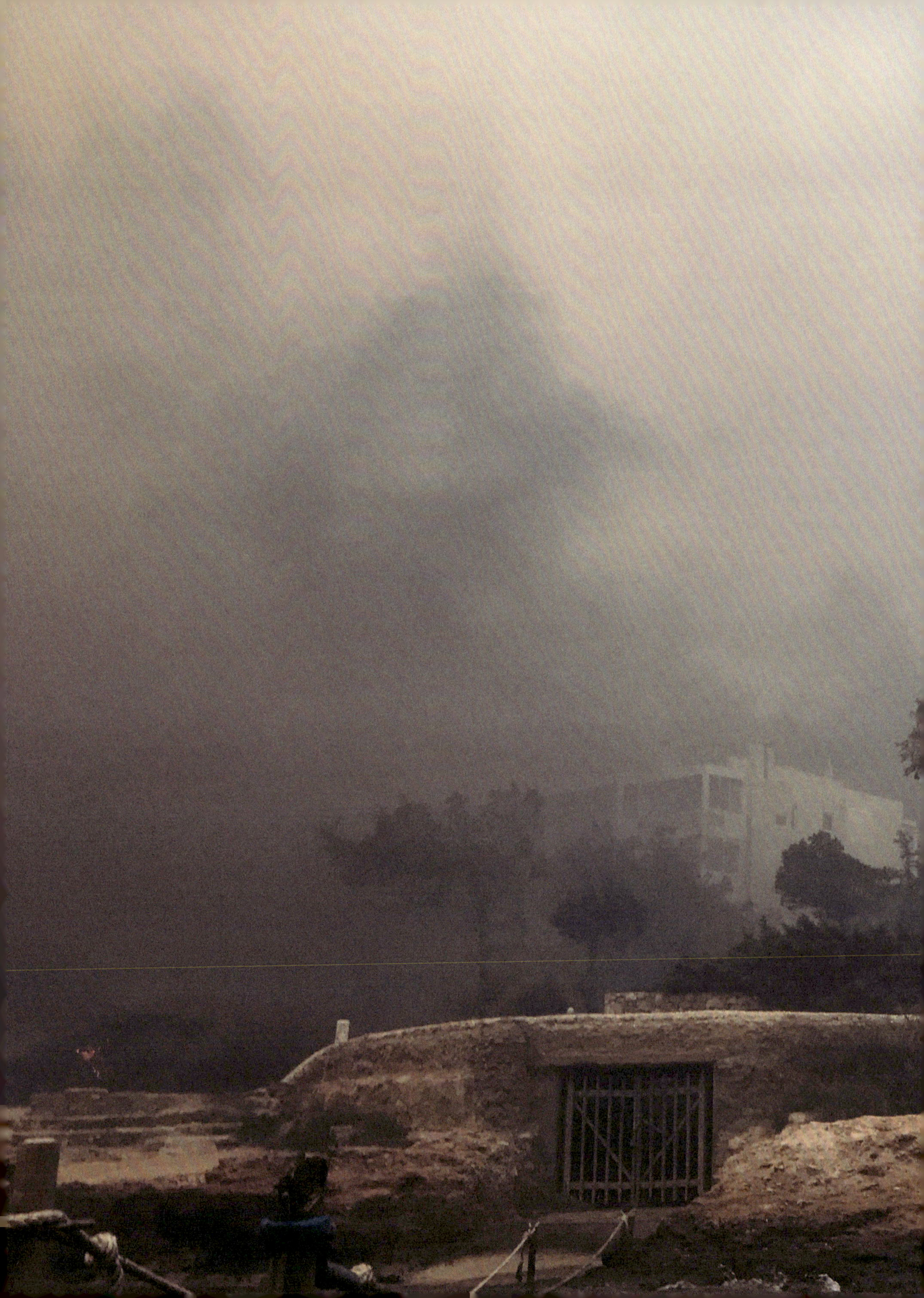

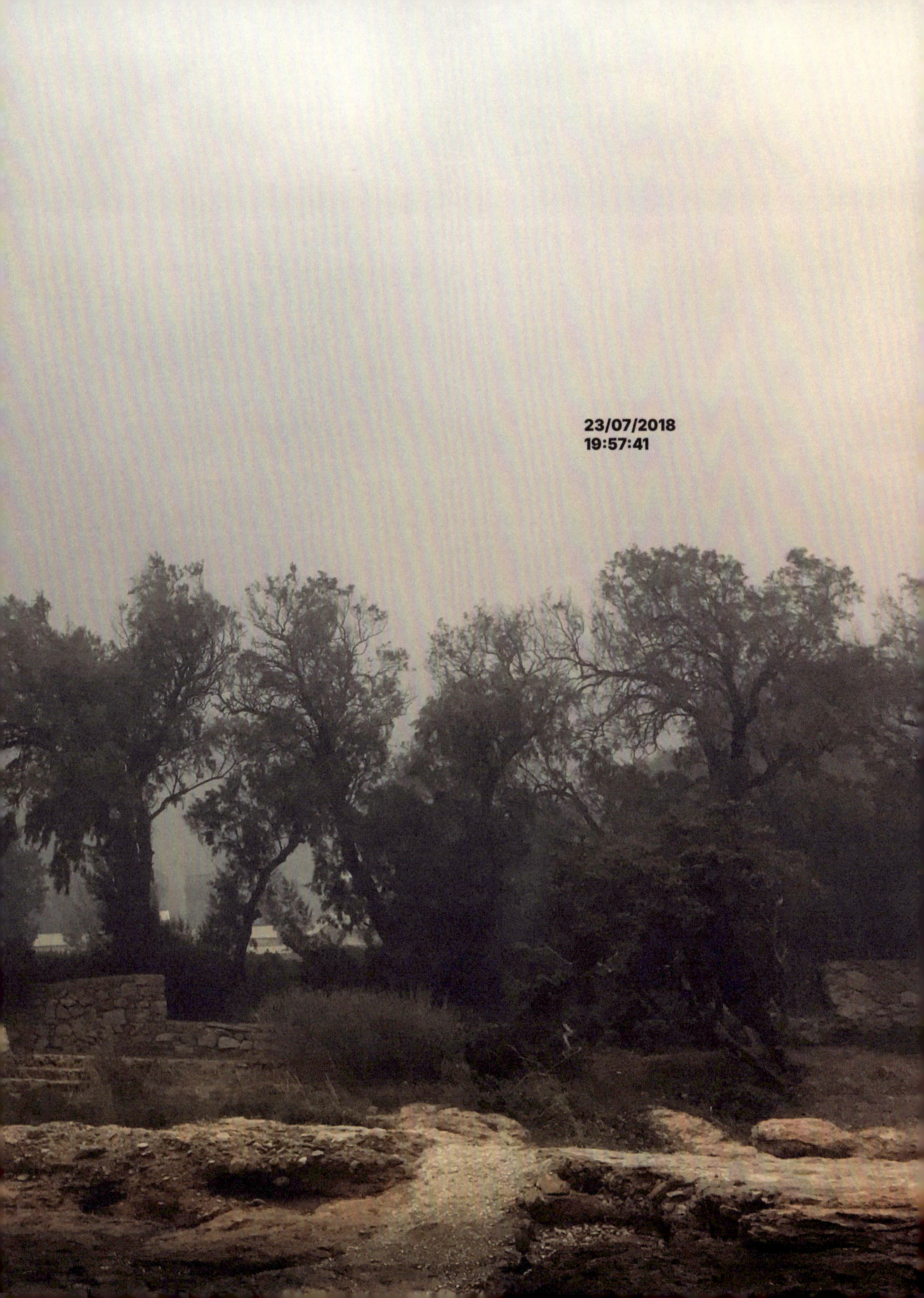
23/07/2018
19:57:41

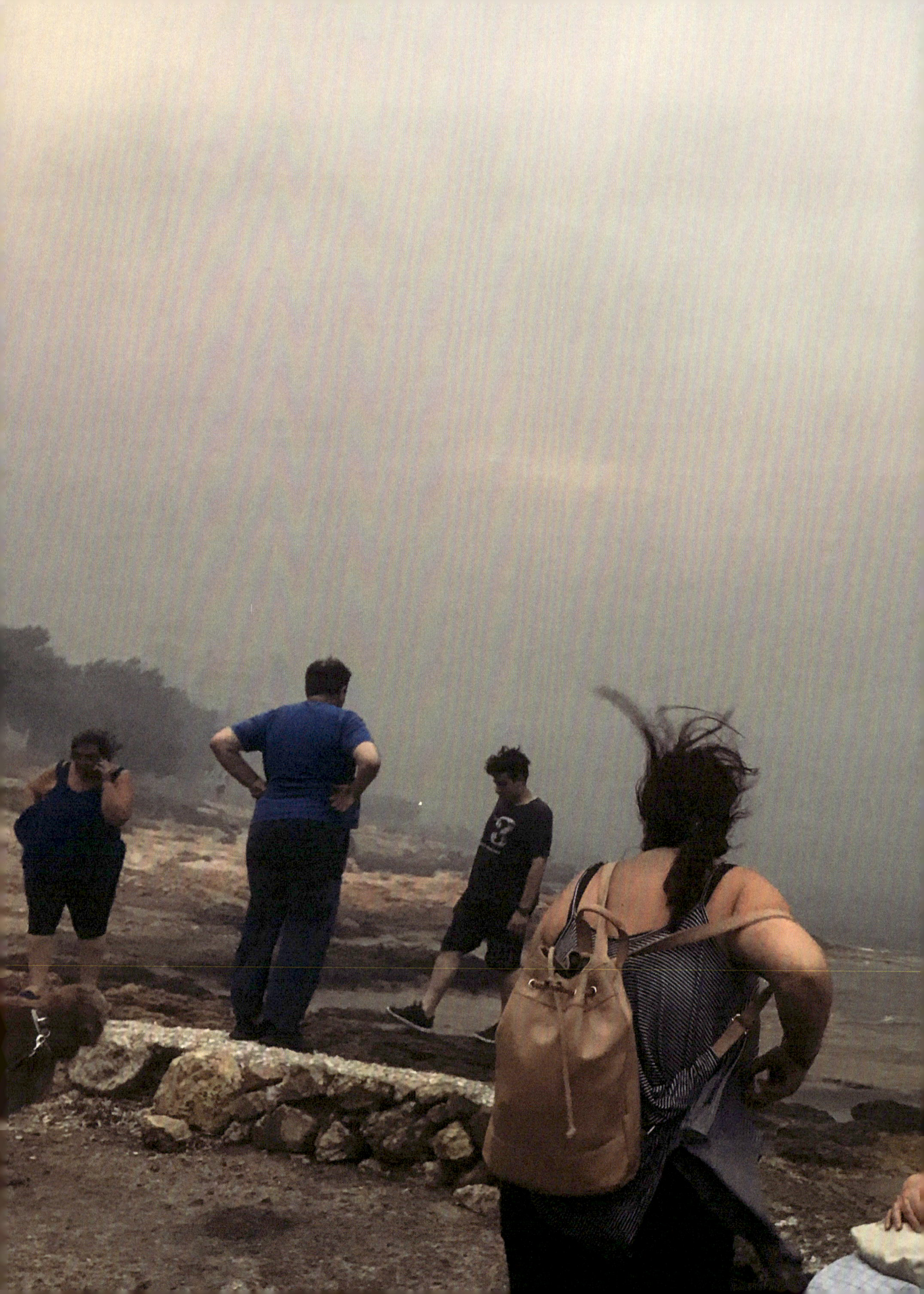

23/07/2018
20:20:41

23/07/2018
20:21:03

There is thick black smog, the crackling of trees going up in flames and the sound of things blowing up. 'We should swim,' says my mother. The only thought that keeps circling in my head is who—my mother, who uses a walking stick and can't breathe, or my 3-year-old—will I have to let drown first? Who do I choose to give up on first? I am not leaving her again. I just say, 'no, we are not going to swim.'

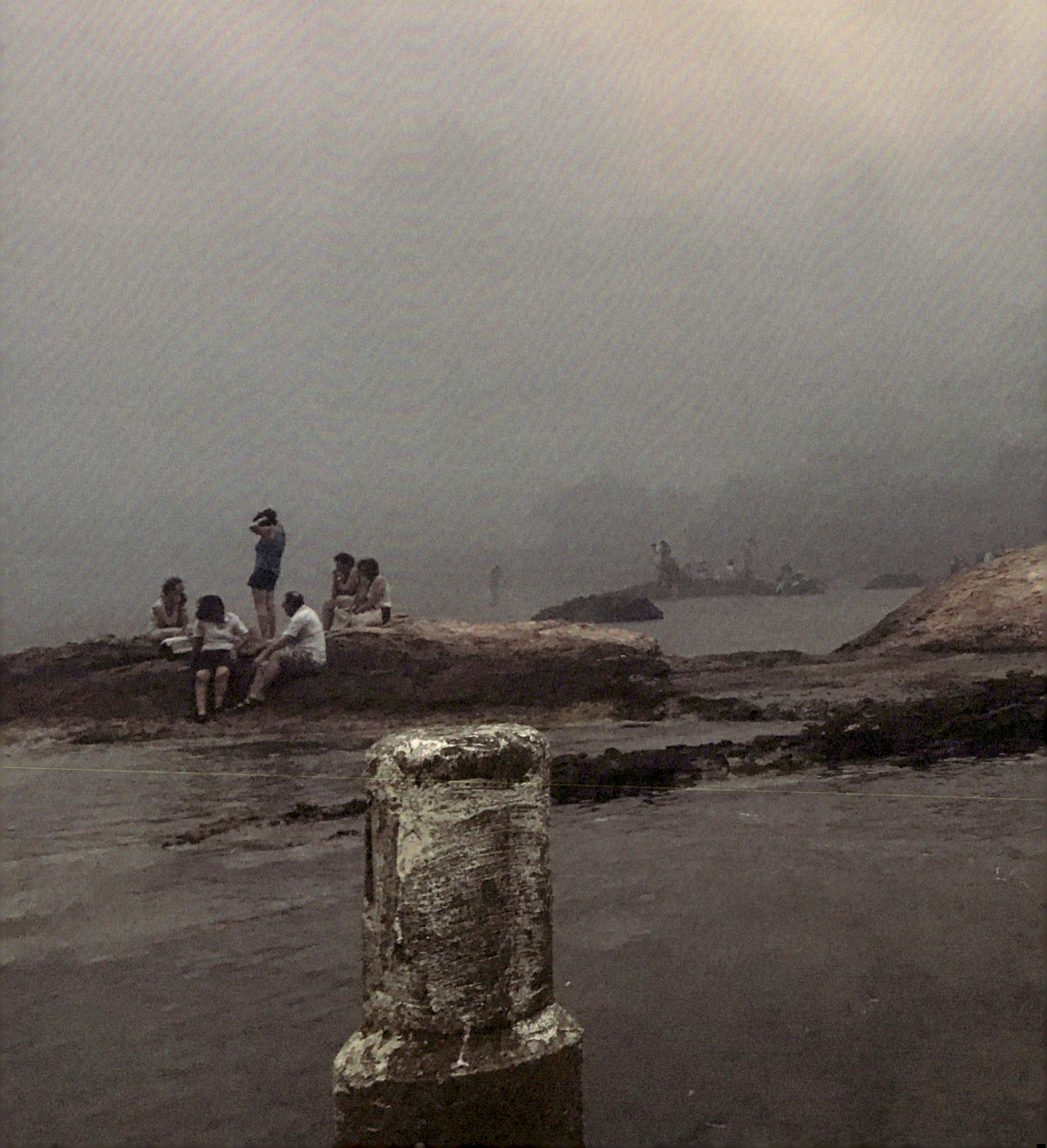

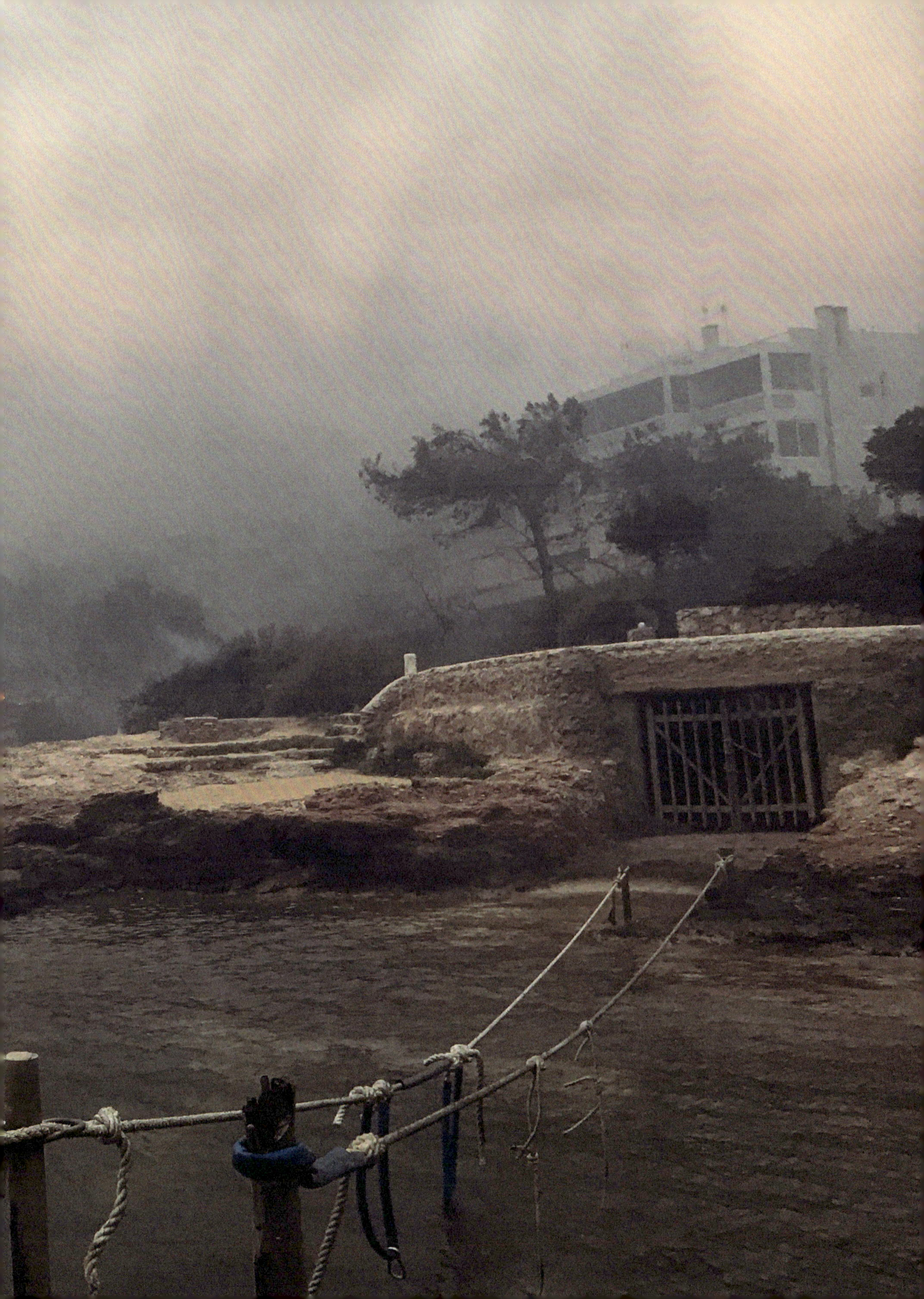

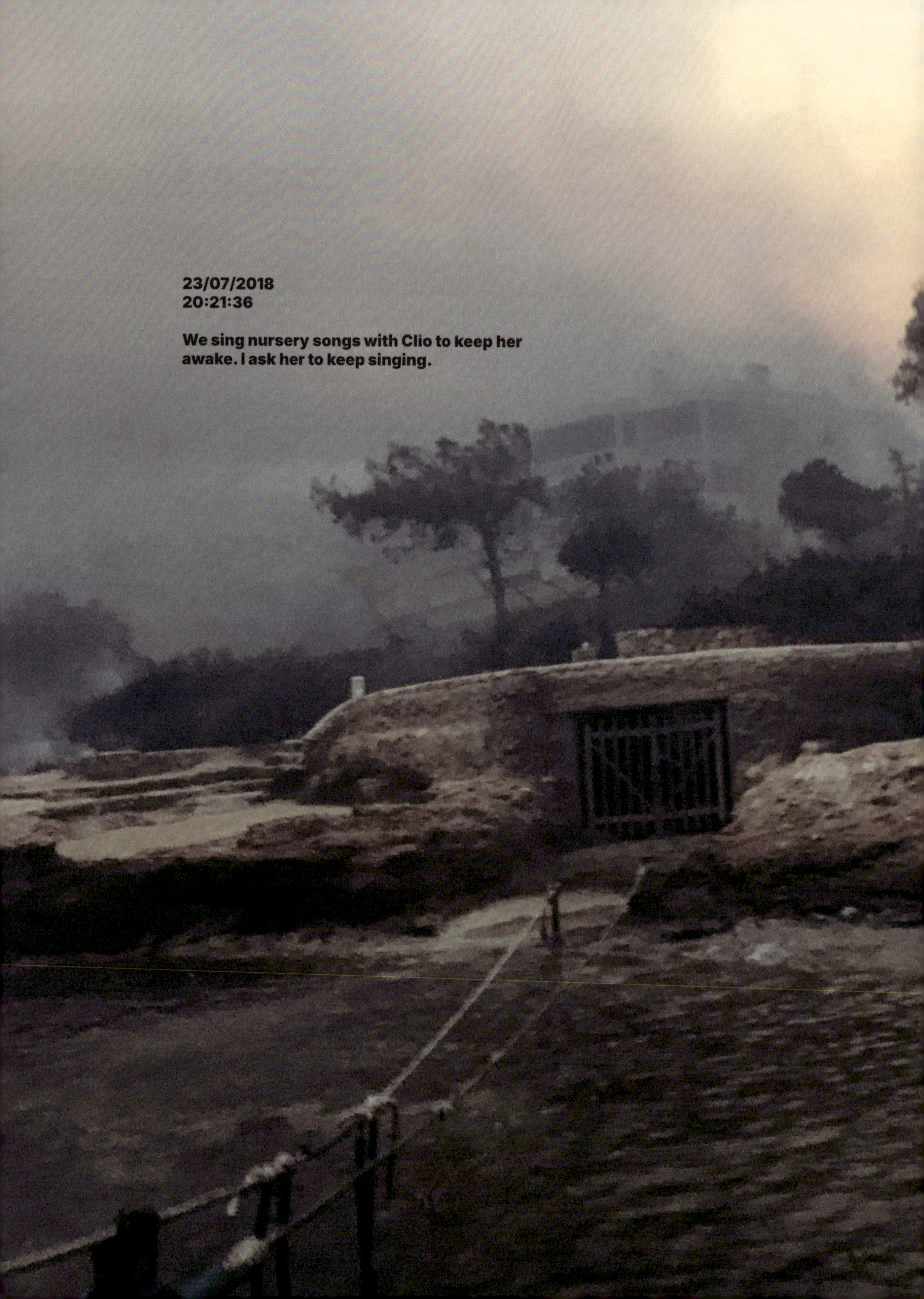

23/07/2018
20:21:36

We sing nursery songs with Clio to keep her awake. I ask her to keep singing.

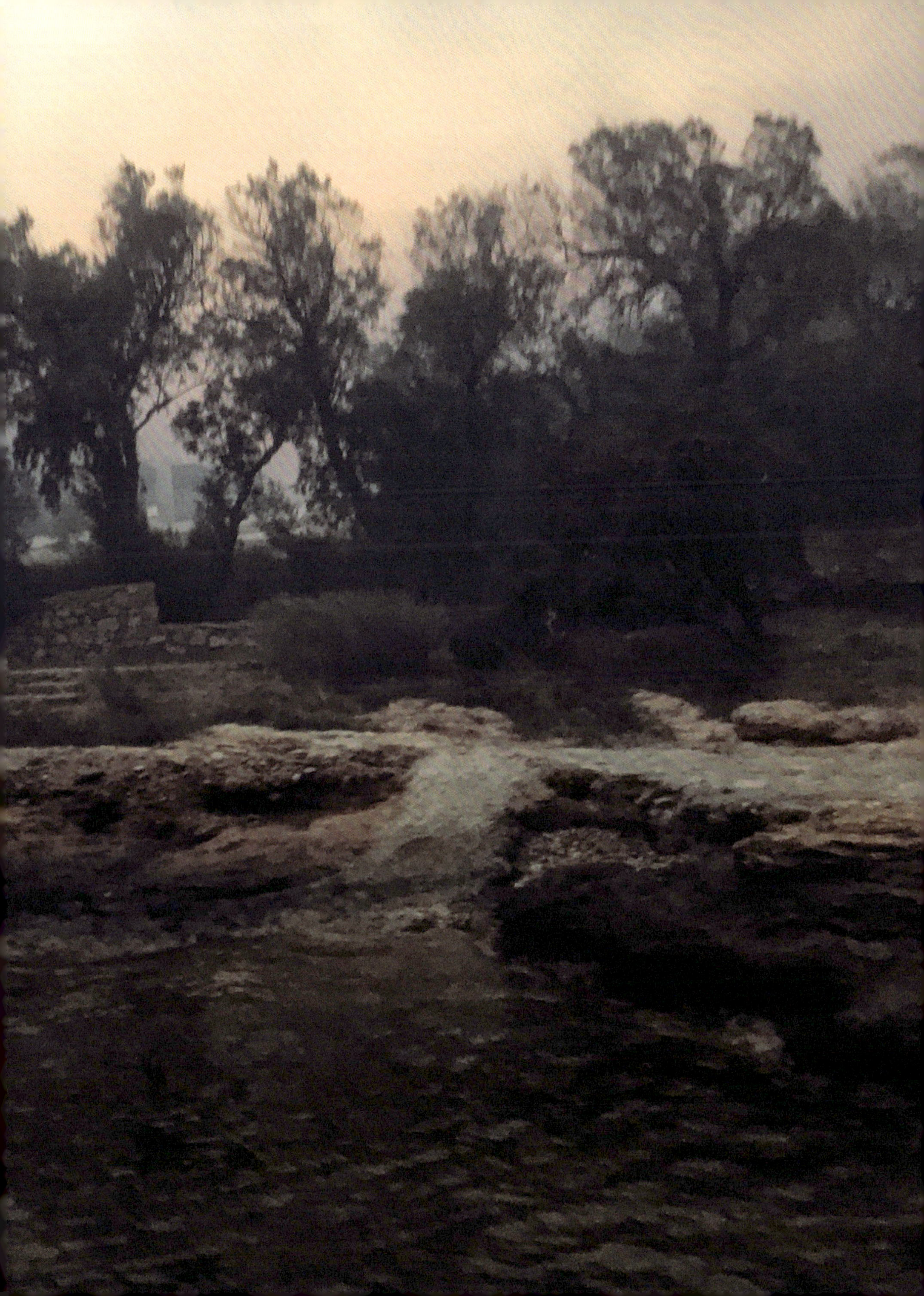

23/07/2018
20:22:39

We take a picture of us. I fear we will not make it.
I am hoping the phone will.

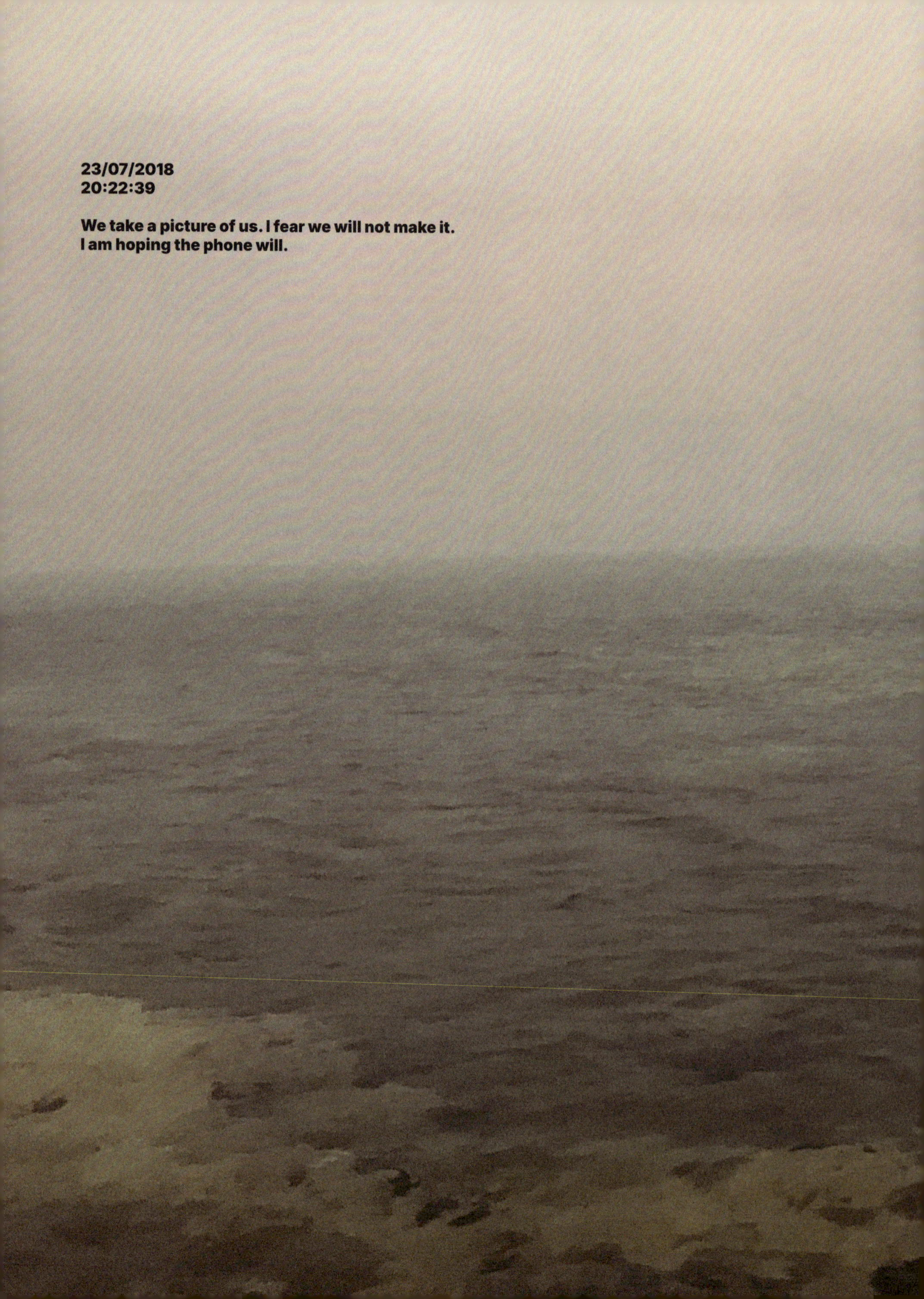

23/07/2018
20:25:46

It's been two hours, maybe more since we first got in the water. The phone has stopped ringing, no one can reach us, and we can't reach anyone. Clio is still relatively calm. She's wet to the bone, shivering, and falling in and out of consciousness. Is she tired, hypothermic or have her lungs given up? I need to get her out but there is nowhere to go.

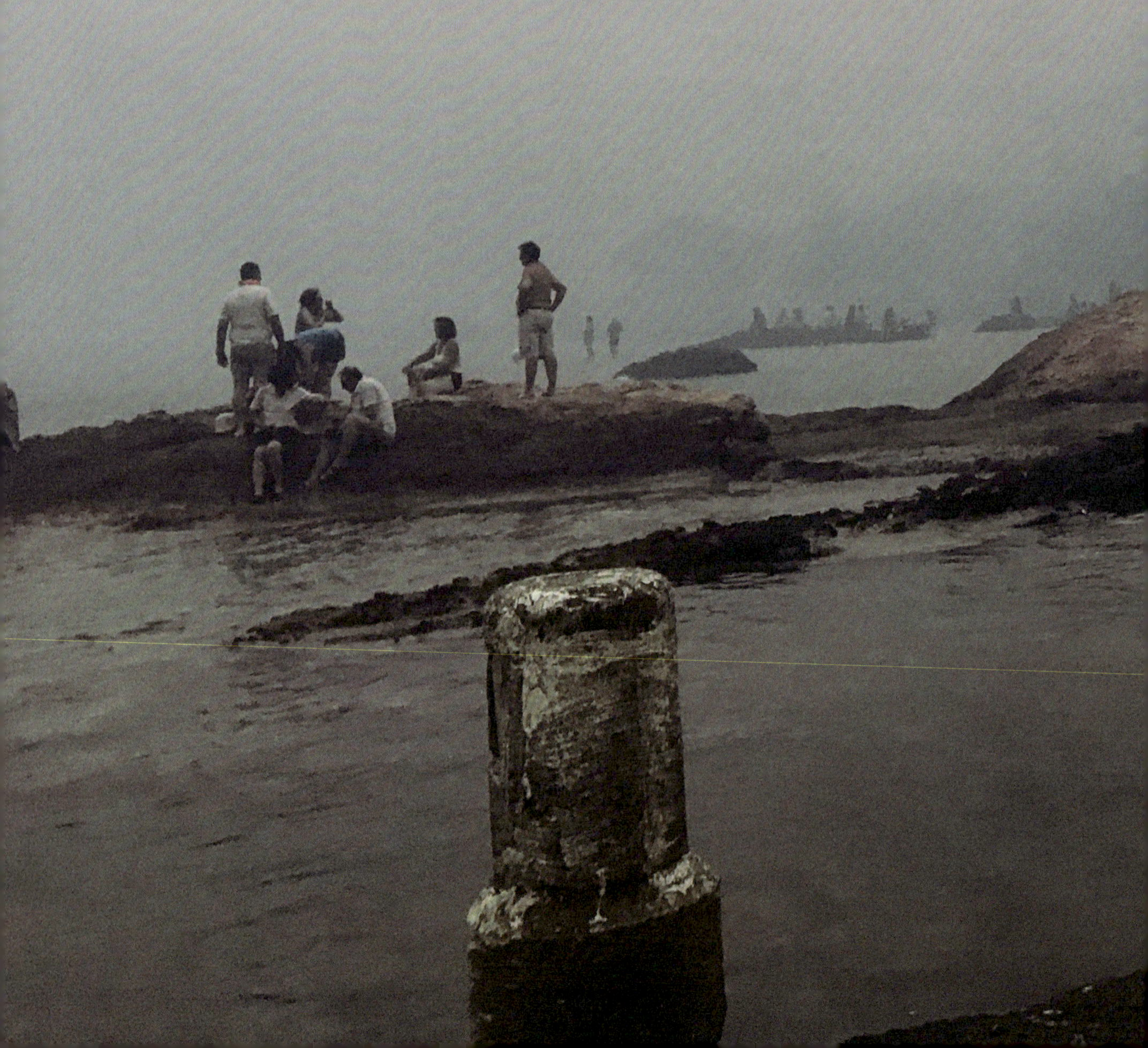

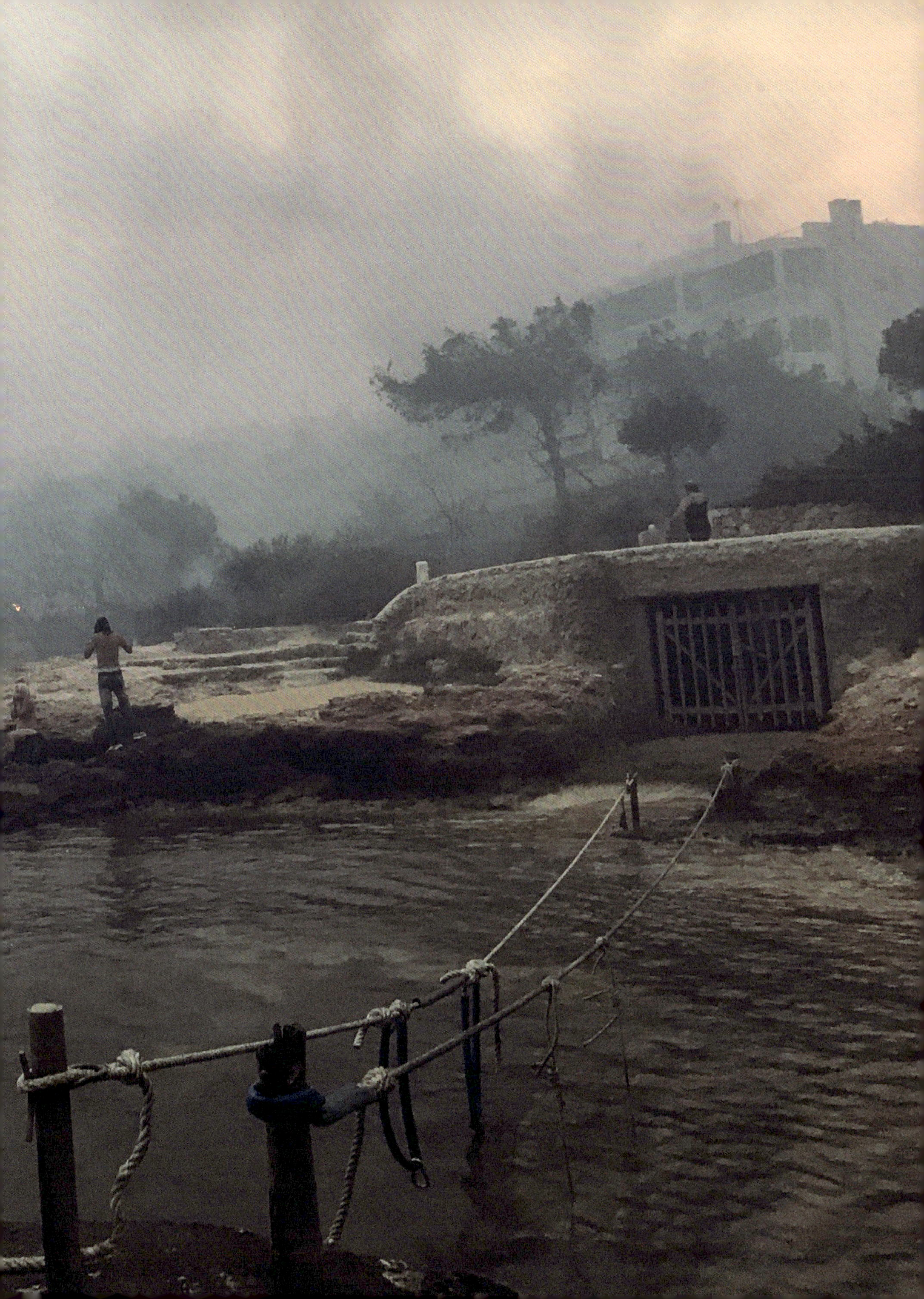

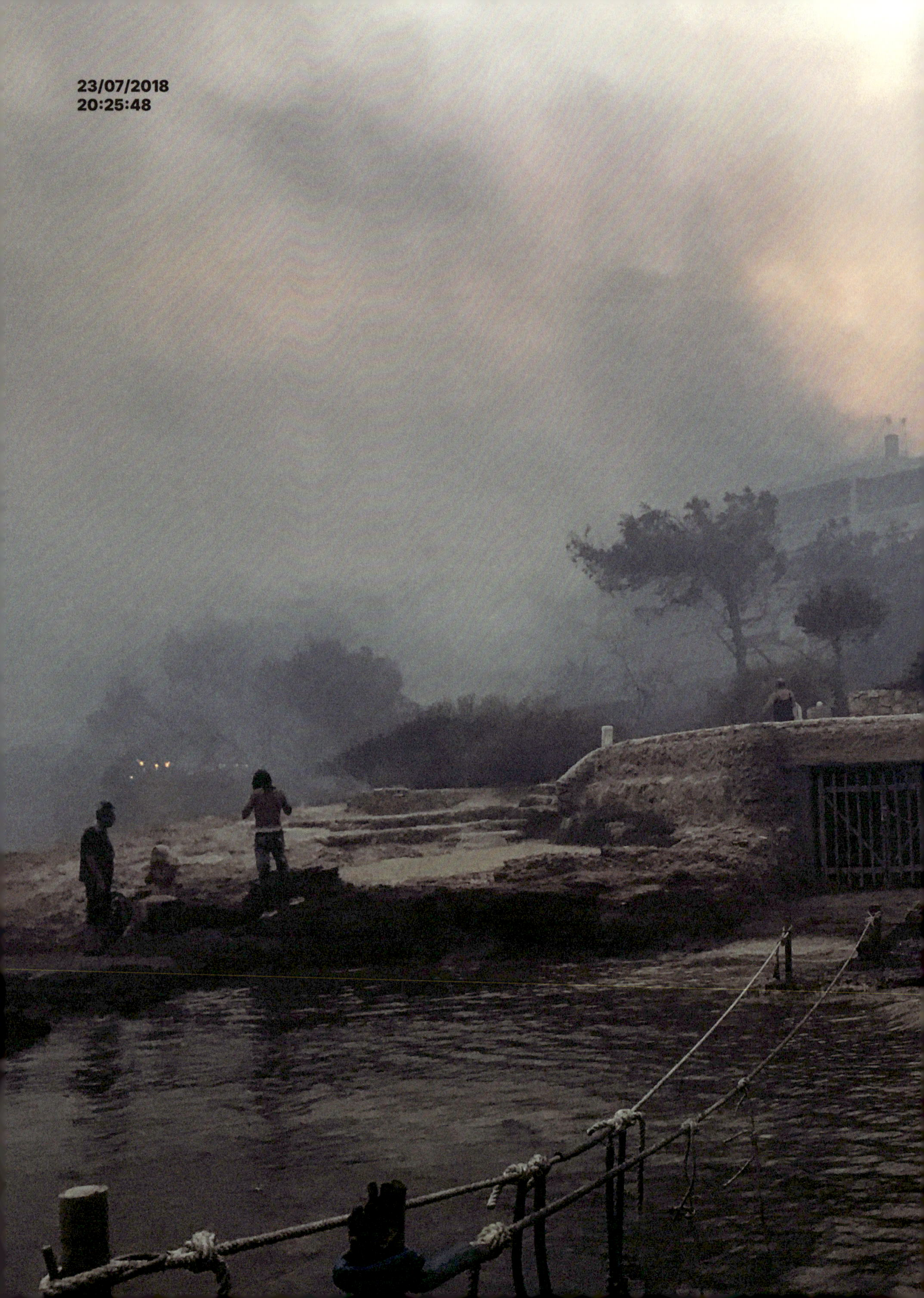
23/07/2018
20:25:48

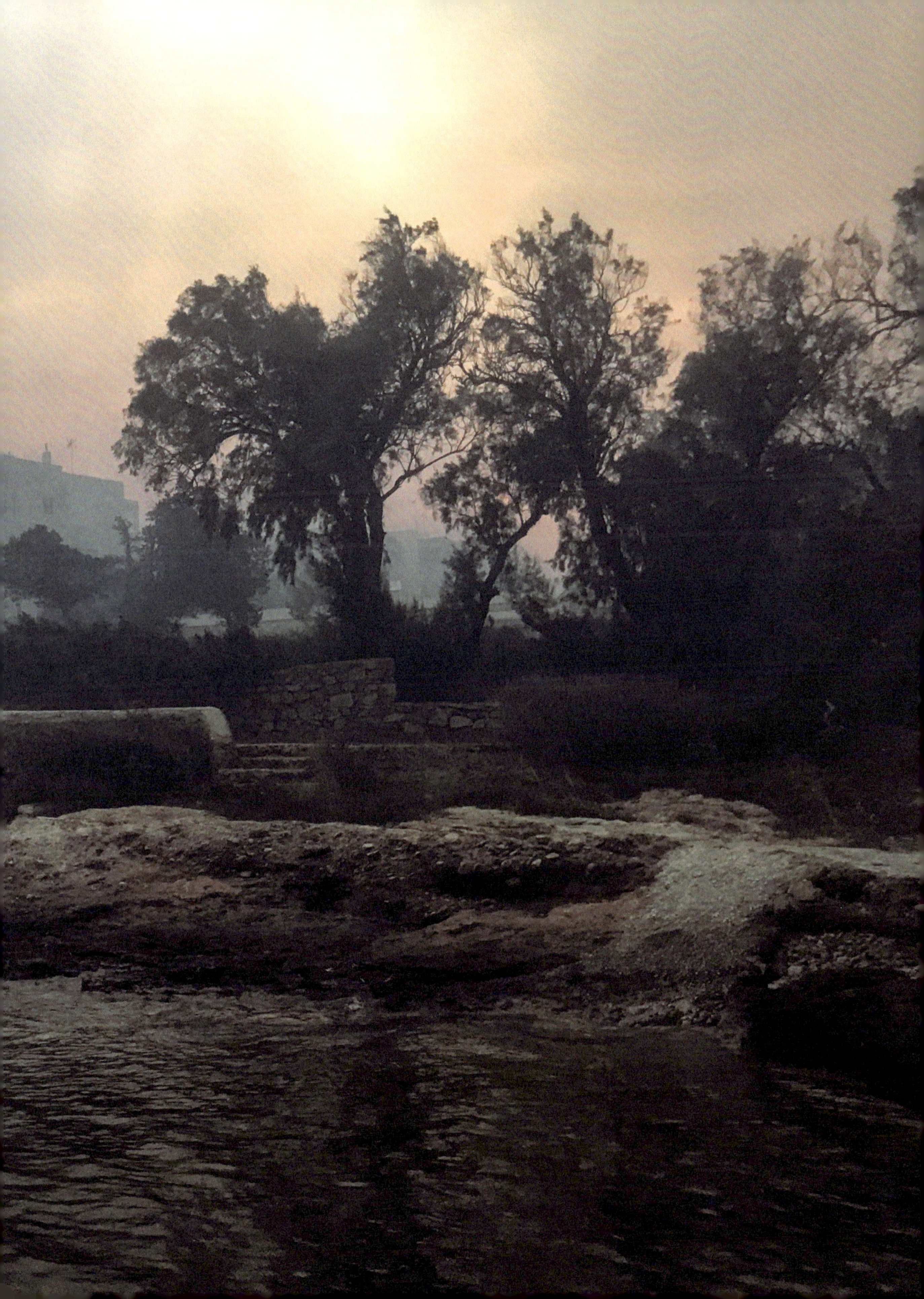

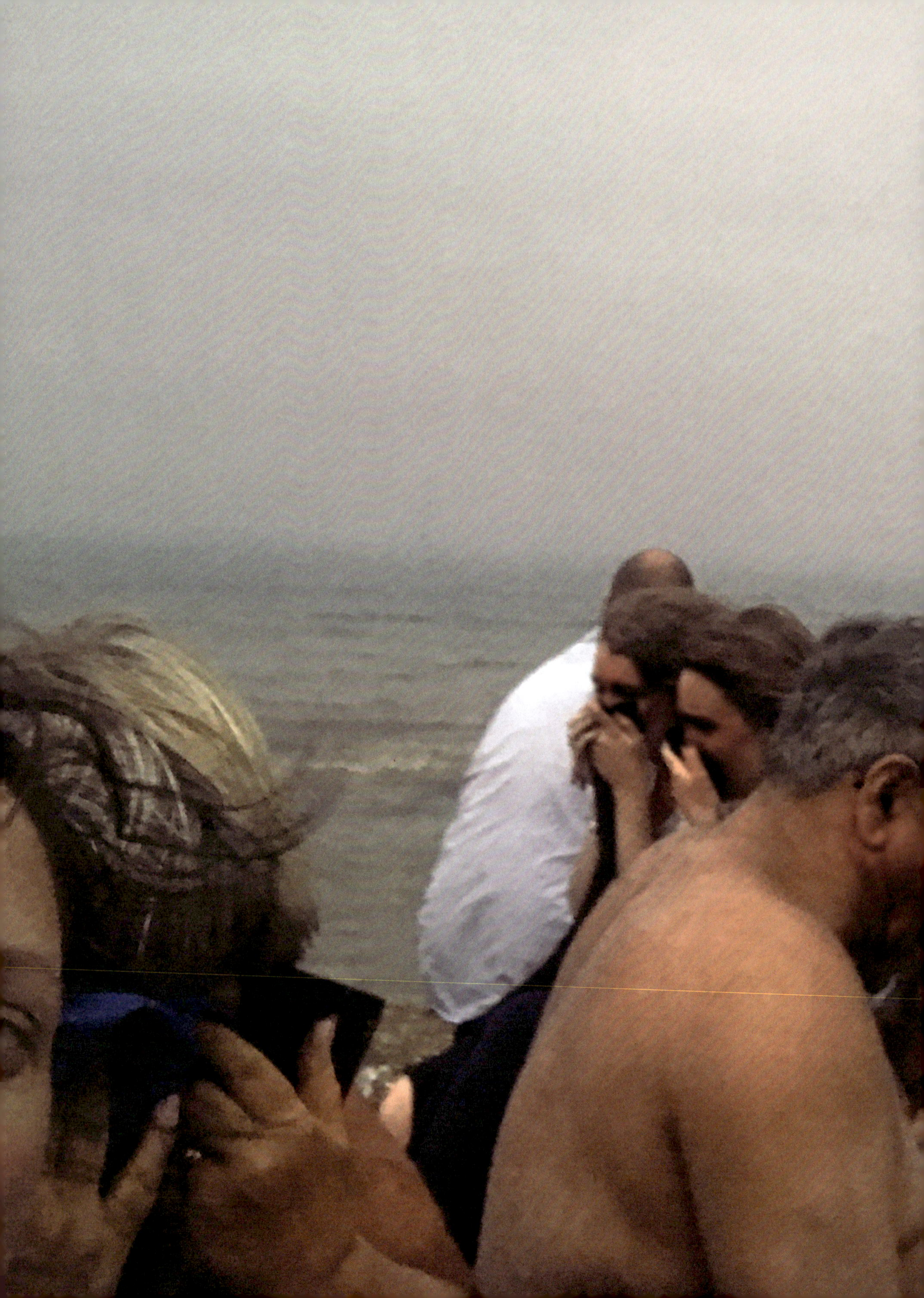

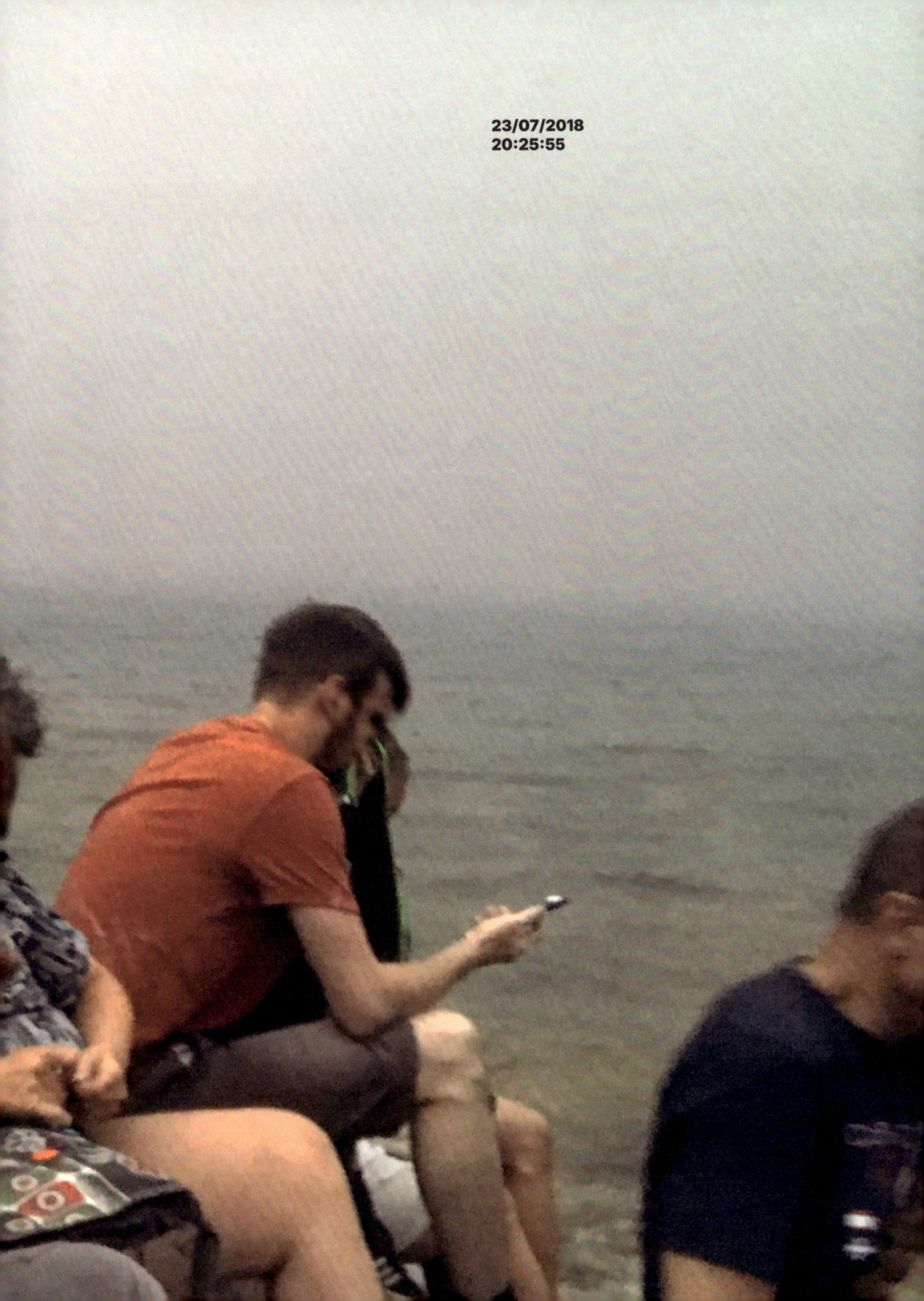
23/07/2018
20:25:55

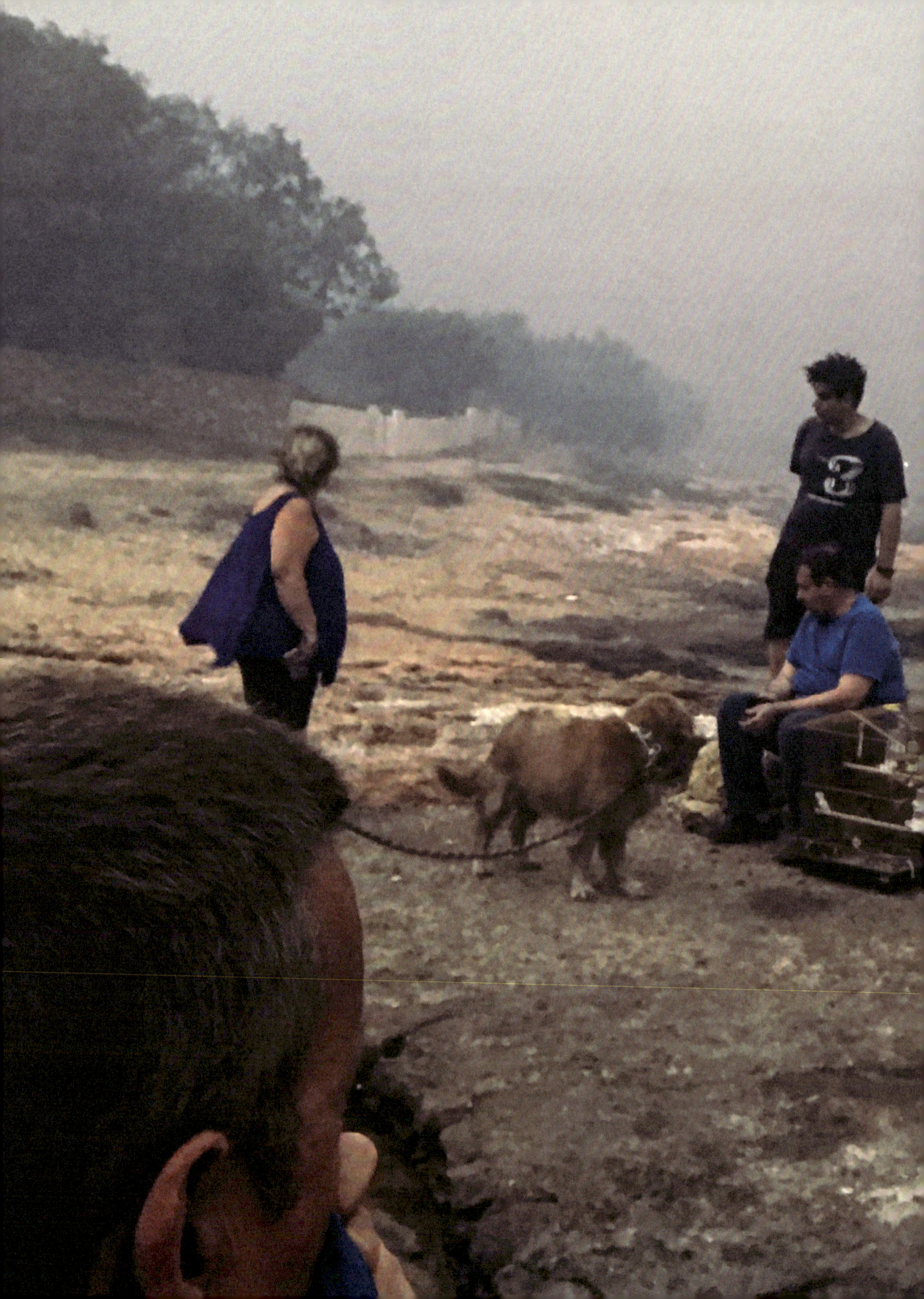

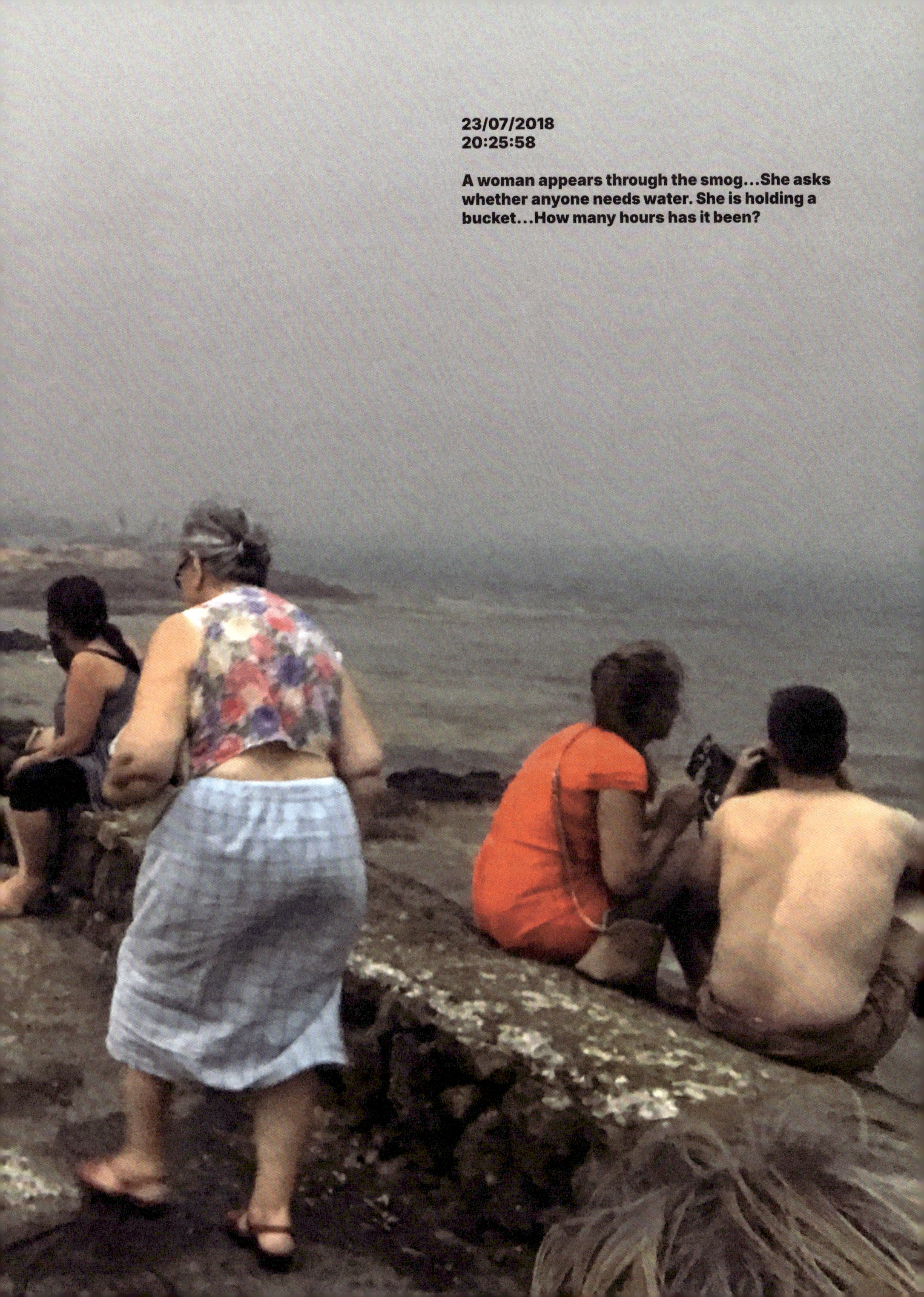

23/07/2018
20:25:58

A woman appears through the smog...She asks whether anyone needs water. She is holding a bucket...How many hours has it been?

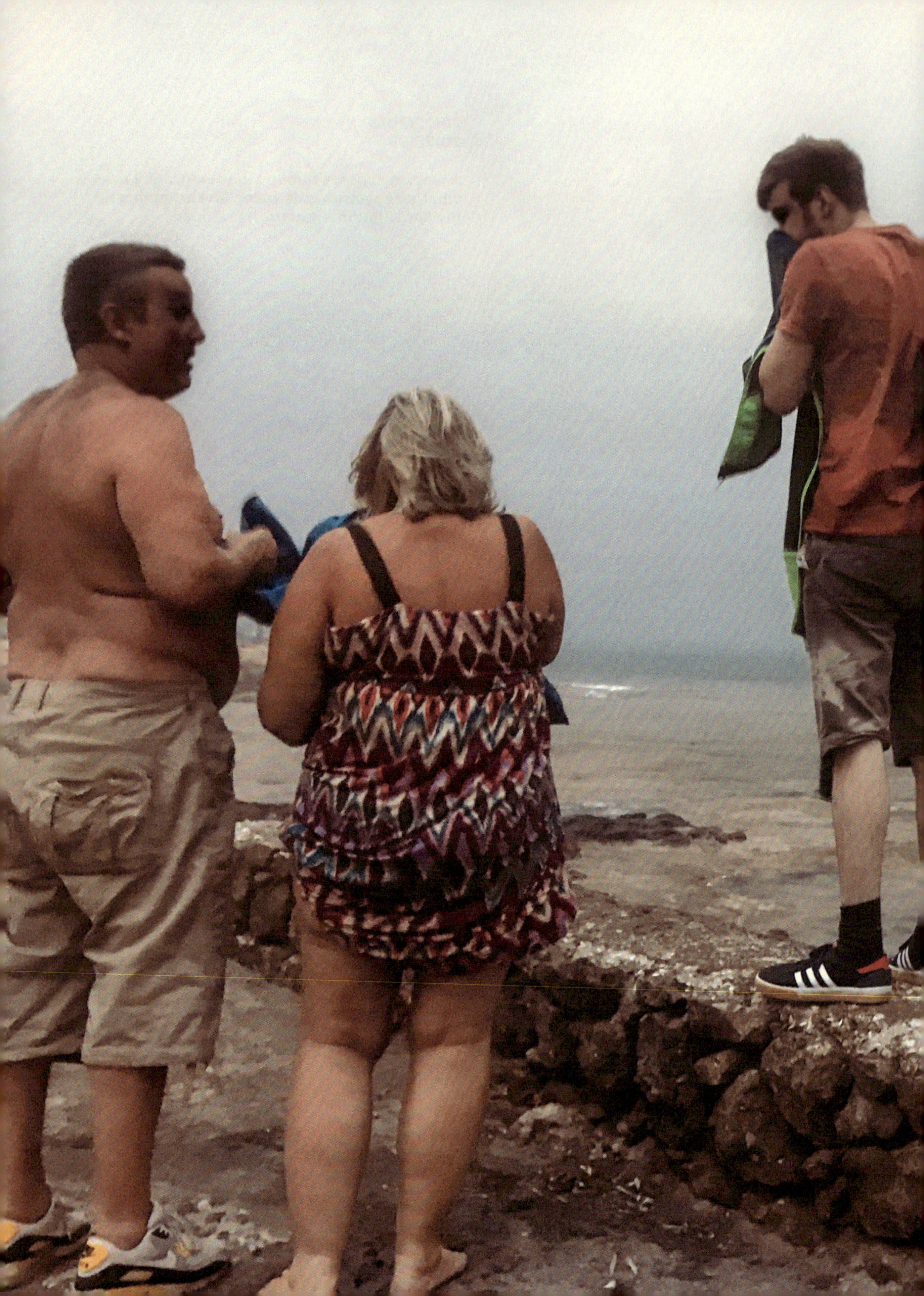

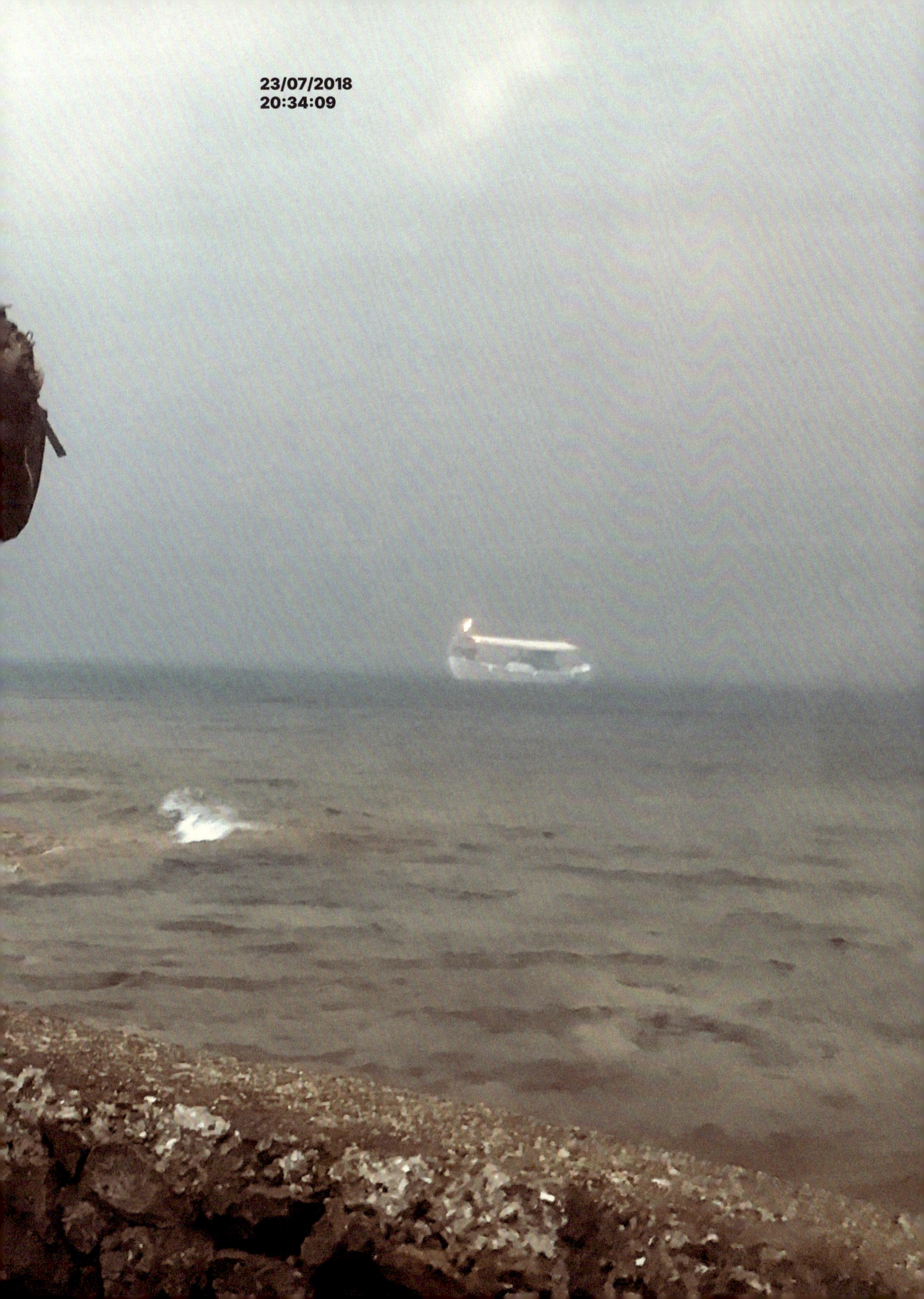
23/07/2018
20:34:09

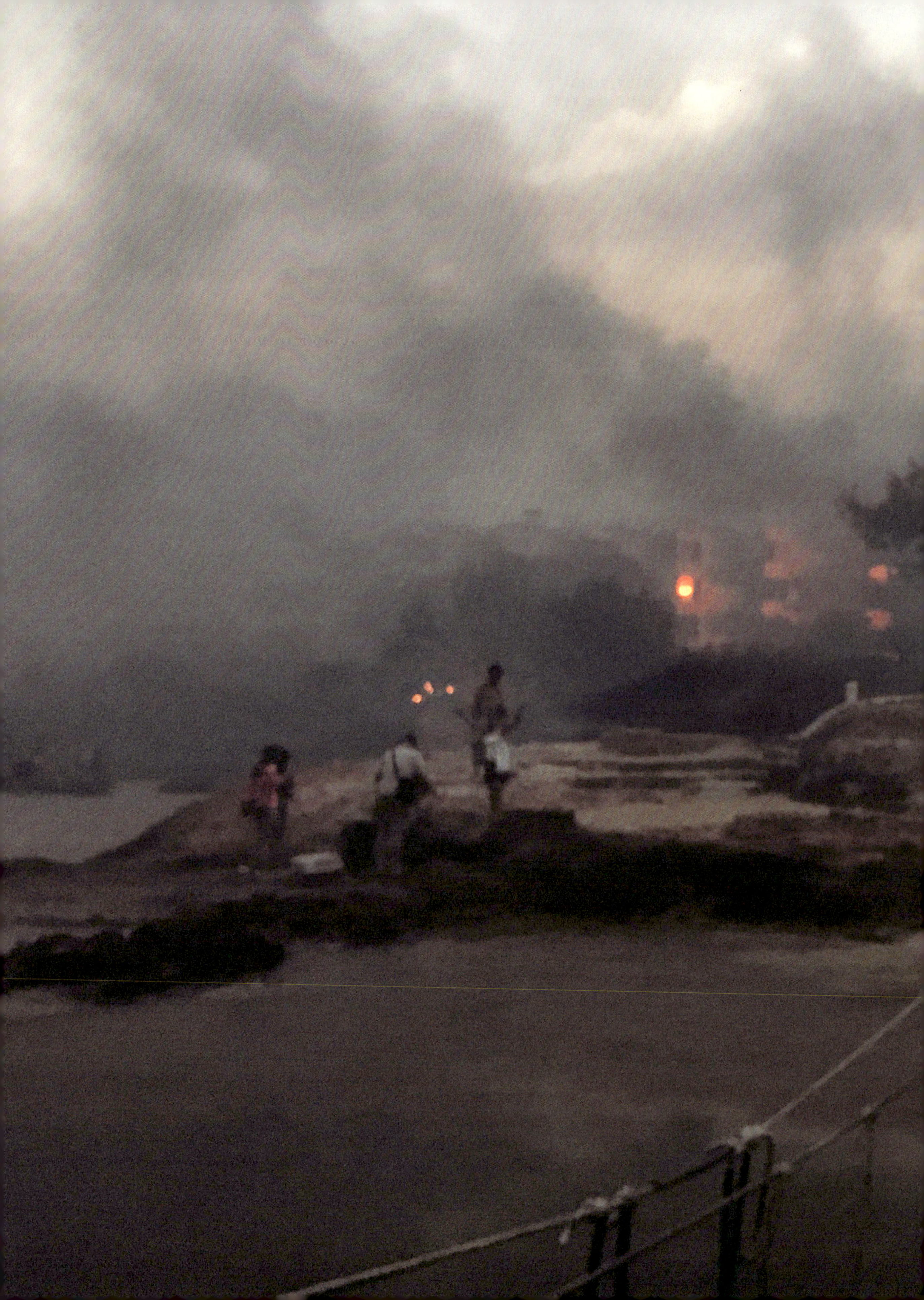

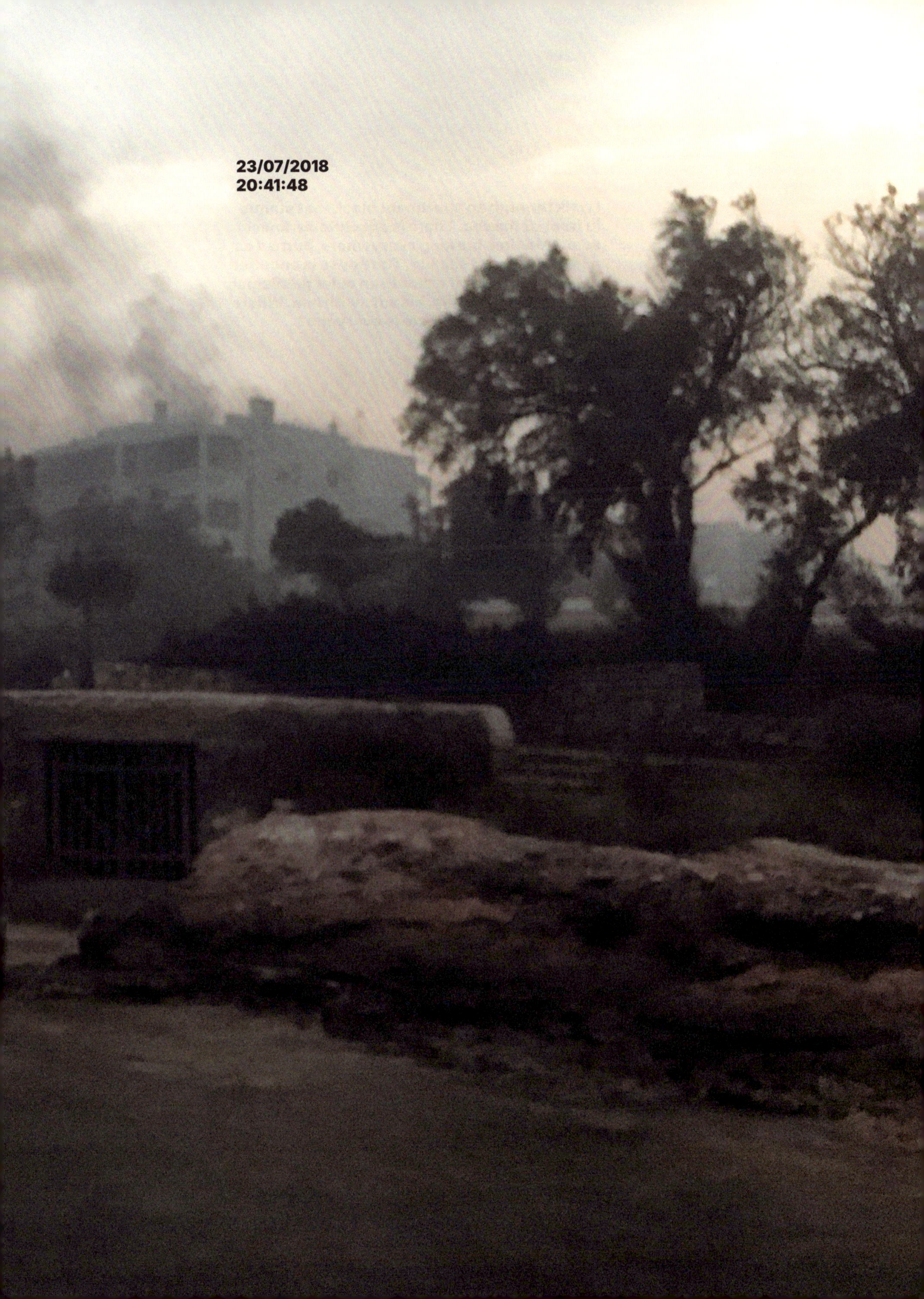
23/07/2018
20:41:48

23/07/2018
20:50:59

I walk through an apartment block that stands in front of the sea. There is absolute darkness, scattered fires burning everywhere, burned cars and rubble—and silence. The house is engulfed in flames. We bring water to an older gentleman who has been burned, his wife with him. Where is the ambulance? Where is everyone?

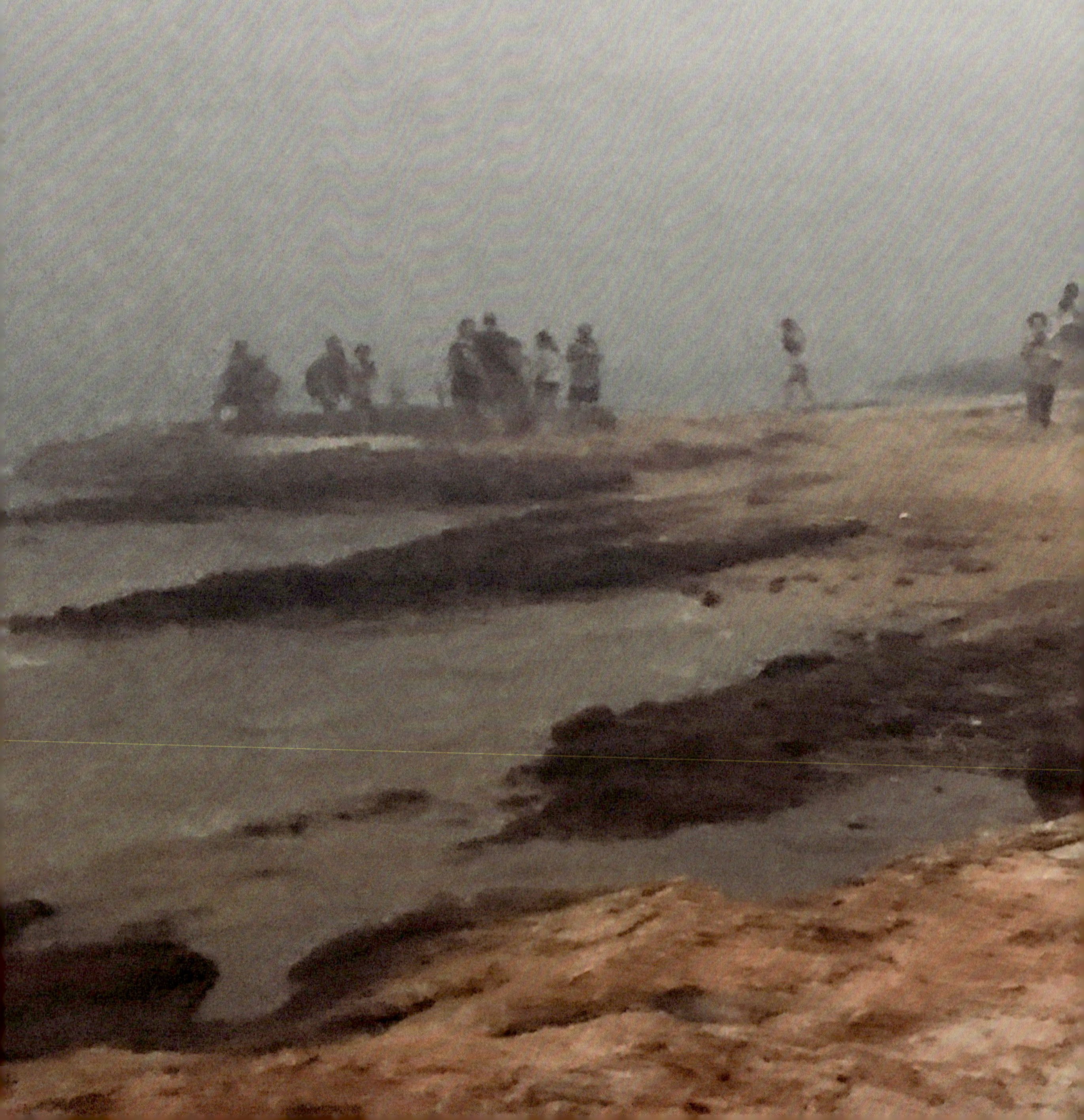

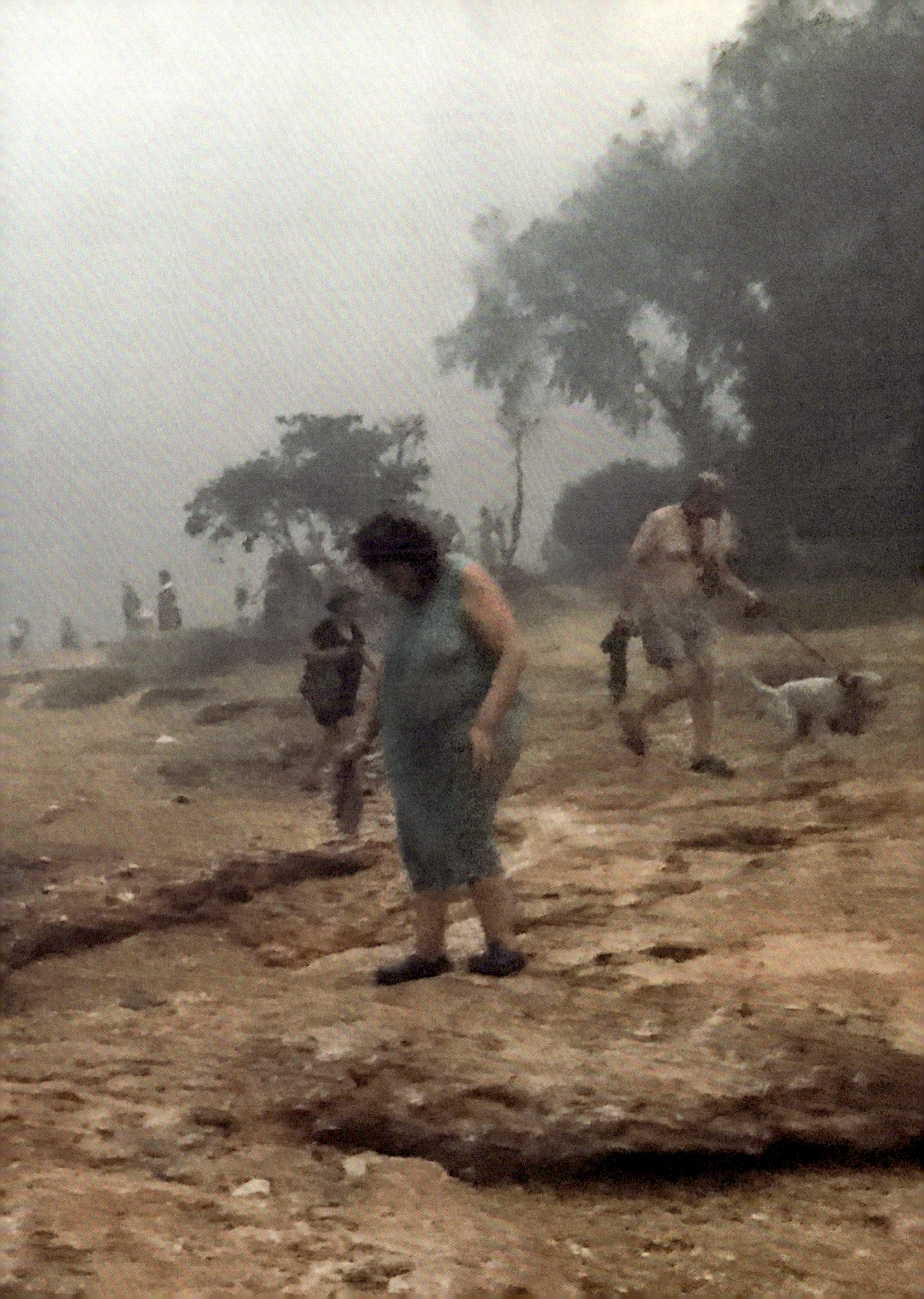

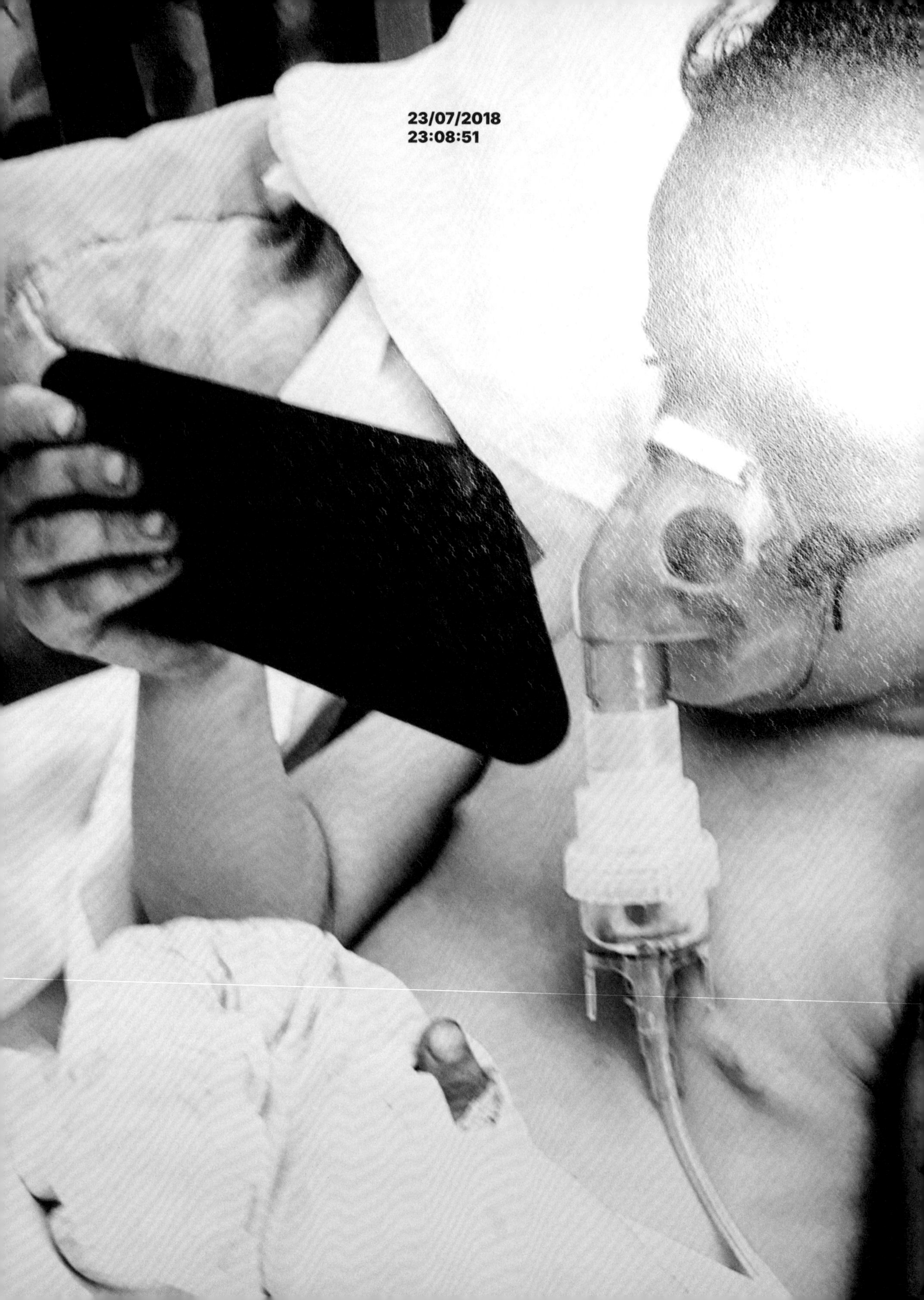
23/07/2018
23:08:51

104 Victims

Location of Death

74 At the Location where the fire took place.
21 In a hospital or medical facility where they had been transferred with injuries or in their homes the following days or months
9 Drowned at sea

Gender

55 Female
49 Male

Age

60 Older than 60-years-old
44 Younger than 60-years-old,
8 of which younger than 10-years-old

74 Victims that died on location of which

33 In Mati, mostly on Leoforos Poseidonos
31 In Kokkino Limanaki, of which 26 in single plot on Argyra Akti
7 In N. Voutzas
2 In the port of Rafina
1 In an ambulance

70 In open spaces i.e. streets, plots, campgrounds, beach, etc
3 In a house, garage etc
1 In a car

Information from the Report of Technical Advisor and Judicial Expert, Andrianos Gourbatsis.

Y'

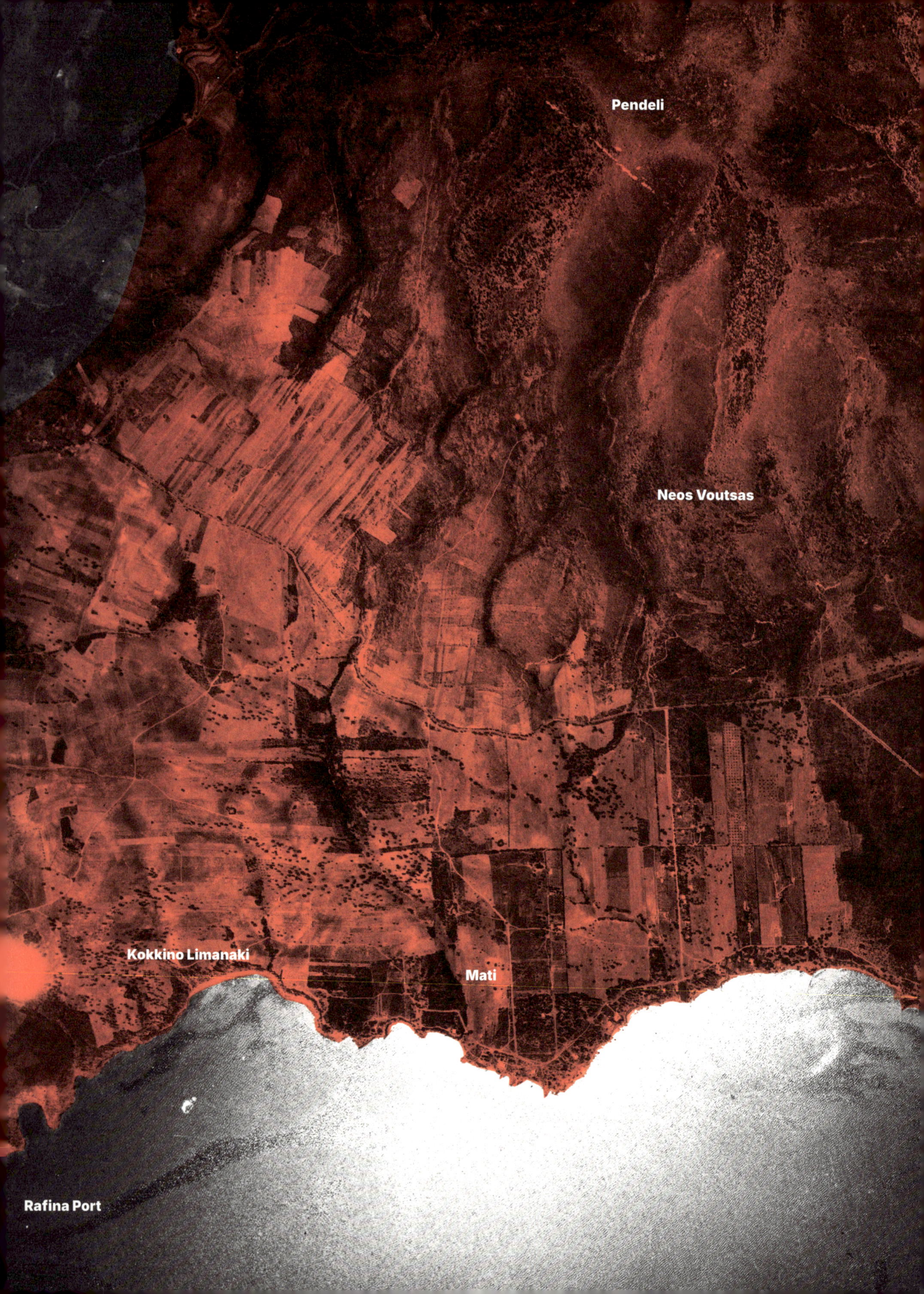
Pendeli
Neos Voutsas
Kokkino Limanaki
Mati
Rafina Port

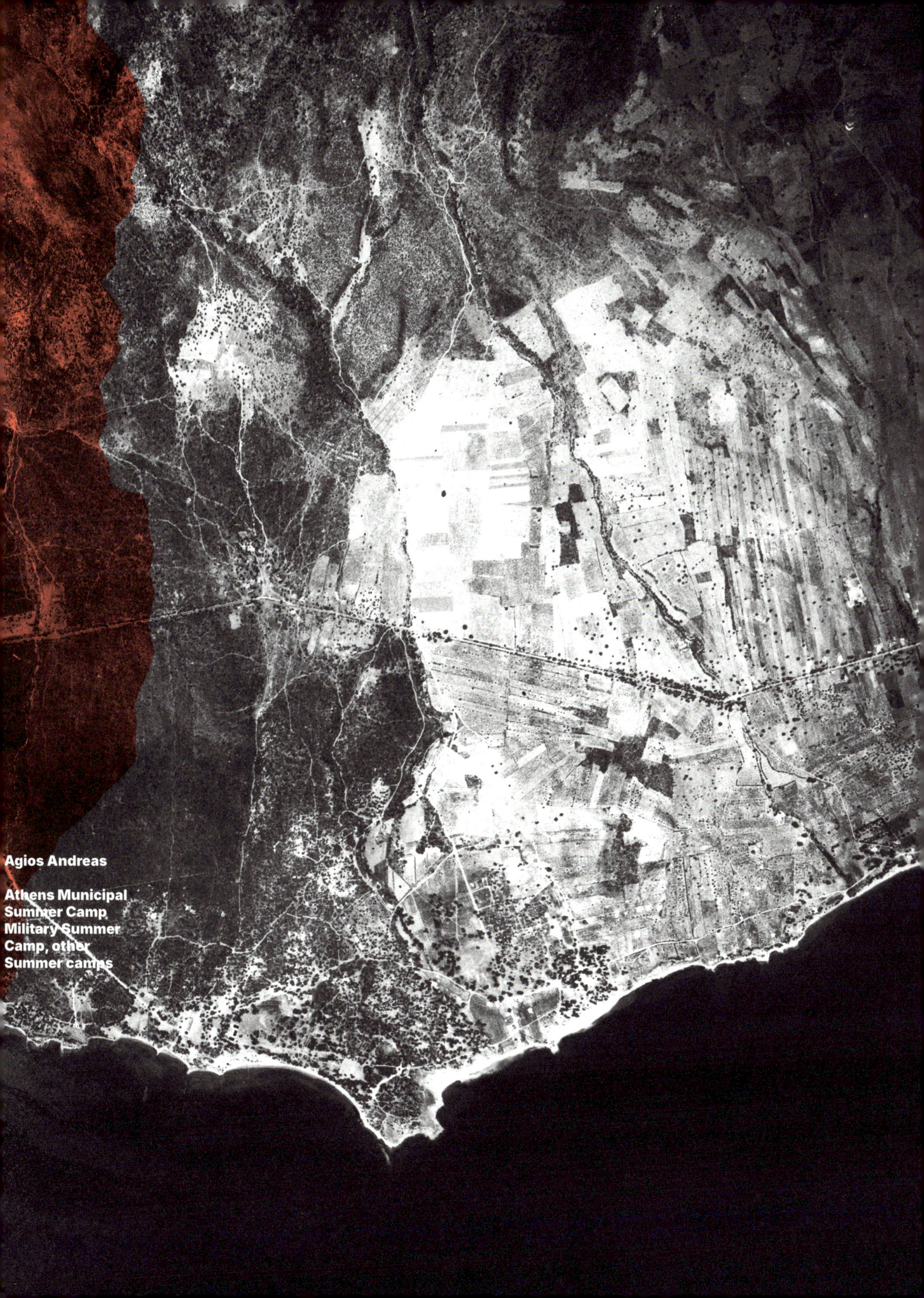
Agios Andreas
Athens Municipal
Summer Camp
Military Summer
Camp, other
Summer camps

23 July 2018

Fire warning level 4—Very high

No alternative stationing of aircraft on 22 July, although there are severe wind warnings that could possibly affect aircraft taking off.

No request on time to the Airforce for Aerial Supervision of the Area of Attica with a water carrier Aircraft. The request to the Airforce was not sent before 8:00 (according to protocol), and therefore was denied.

The request was sent at 9:59.

A small volunteer plane (CL72 non water carrying) from an Airclub accepts the request to supervise the area. It performs two patroling flights 14:01–15:55 and 16:16–18:08, but does not spot the fire.

There are fifteen aircraft (including three non water carrying) available to patrol in Attica that day.

12:03 A fire breaks out in location 'Aeras' in the Geraneia Mountains (Kineta). A forest area of approximately 55,680 acres is destroyed.

Location 'Aeras' is 94.4km from Ntaou Pendeli and 67.2km from the centre of Athens.

12:48 Vehicle PS 3823 with call code (MAKRI–15), patrol fire engine (water –5000 litres), of the Nea Makris Fire Department is dispatched by the ESKE (Central Operations Coordination Centre) /199-SEKYPS (Operations Coordination Centre for the Fire Department) for assistance to the Kinetta fire. The time of its arrival in the area is not recorded.

Vehicle PS 3823 is stationed at the first entrance of Neos Voutzas, on Marathonos Ave.

Note: Vehicle PS 3823 is one of the two patrol vehicles that would have been closest to the ignition point of the fire of Ntaou Pendelis. If it had not been moved to Kinetta, it would have arrived within seven minutes, at 16:48.

13:51 Vehicle PS 1475 with call code (12–30), patrol fire engine (water–2500 litres), of the 12th Fire Station of Athens and stationed at the location 'Patima' of Vrilissia, is despatched by the ESKE/199-SEKYPS, for assistance to the Kinetta fire. Time of arrival 15:26.

15:06 Vehicle PS 3527 with call code (12–15), patrol fire engine (water–5000 litres), of the 12th Fire Station of Athens and based in the Drafi-Dioni area, is despatched by ESKE/199-SEKYPS, for assistance to the fire in Kinetta. Arrival time 16:30.

Note: Vehicle PS 3527 is the other of the two patrol vehicles that would have been closest to the ignition point of the fire of Ntaou Pendeli. If it had not been moved to Kinetta, it would have arrived within five minutes, at 16:46.

16:15–16:30 A fire breaks out in Ntaou Pentelis. Ignition point coordinates: X=494806.858 and Ψ=4209969.075 (Data from the Hellenic Geodetic Reference System 1987/EGSA 87). Ignition point is within an unfenced plot on Androutsou Street. Cause of fire is human negligence.

16:35 An eyewitness sees smoke near the ignition point from the intersection of Plastira and Androutsou streets in Ntaou Pentelis.

16:41 199-SEKYPS is notified about the fire.

16:46 Smoke is visible at a distance of about 70m from the Ignition point and is recorded on CCTV video footage by the security camera of a residence on Ioustinianou street No. 10 at Ntaou Pentelis.

16:47 The first three fire engines (water carriers) are mobilised. There is a delay in the mobilisation by ESKE/199-SEKYPS of the first fire engines by six minutes from the time they are notified 16:41, and by seventeen minutes, from the minimum estimated ignition time of the fire.

The first fire engine (water–2500 litres), PS 3211 of the Nea Makri Fire Station, arrives in Agios Andreas in the wider area of the incident.

16:48 One more fire engine is mobilised.

16:50 A Fire Department's S-64 heavy lift helicopter (sikorsky S-64 Skycrane) receives an order to divert from the Kinetta fire to the Ntaou Pentelis fire.

16:51 Two more fire engines stationed in the 12th Fire Station of Athens, are mobilised.

16:55 The Commander of the 12th Fire Department of Athens (codename 12–10) arrives at the incident.

16:56 Arrival of the first fire engine at Ntaou Pentelis. Efforts to suppress the fire begin.

17:00 The Commander of the Fire Department of Nea Makri (codename ARIS NEAS MAKRIS) arrives at Neos Voutzas.

Evacuation order is given to the Athens Municipality Camps of Agios Andreas and Zoumperi.

The Fire Department issues a Press Release: 'The Civil Protection Deputy Minister Mr. Toskas, together with the Secretary General of Civil Protection Mr. Kapakis, the Chief and the Deputy Chiefs of the Fire Department are at the ESKE and are coordinating the firefighting operation from the first moment.'

17:04 One more fire engine arrives at Ntaou Pentelis. There are currently two fire engine water carriers at the incident at Ntaou Pentelis.

16:47

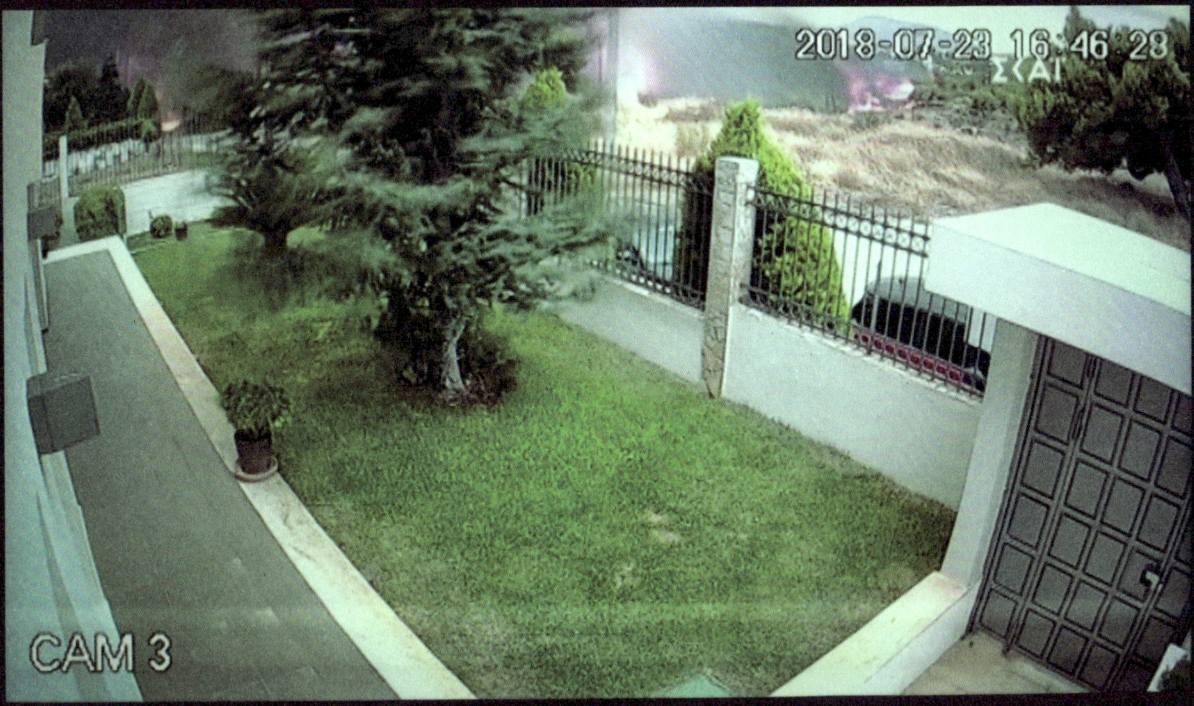

17:05 One more fire engine arrives at Ntaou Pentelis. There are currently three fire engine water carriers at the incident at Ntaou Pentelis.

17:06 One more fire engine arrives at Ntaou Pentelis. There are currently four fire engine water carriers at the incident at Ntaou Pentelis.

17:10 First aircraft—helicopter S64N–154 arrives at the incident. It informs the leadership (Operational and Political) of the Fire Department and the Director of ESKE that the fire is close to homes and within their yards.

There are currently eight fire engine water carriers at the incident at Ntaou Pentelis.

17:15 The original single fire front escapes the Ntaou Pentelis area at the location of the Pantocrator Monastery and splits in two. Both new fire fronts are moving at high speed. One towards Kallitechnoupoli—Kokkino Limanaki and the other towards N. Voutzas—Mati.

17:17 Coordinating helicopter FLOGA 1 arrives at the incident.

17:18 The Commander of the Athens Fire Department (codename ARIS ATHENS) arrives at Ntaou Penteli.

17:20 FLOGA 1 informs the Commander of the Athens Fire Department (codename ARIS ATHENS) that the fire is heading towards N. Voutzas.

17:21 The Deputy Commander of the Fire Department (codename ERMIS ATHENS) arrives at Ntaou Pendelis.

The time window for the organised evacuation of citizens is diminishing.

17.30 S64N-154 informs and updates the Leadership (Operational and Political) of the Fire Department and the Director of ESKE again, that the fire is close to homes. It is moving eastwards and will threaten other inhabited areas.

There are currently nine fire engine water carriers at the incident at Ntaou Pentelis.

The time window for the organised Evacuation of residents and visitors to N. Voutza and Mati has closed.

The Fire Department declares a General Alert. All firefighting personnel need to report to their stations.

17:31 FLOGA 1—The coordinating fire department helicopter departs for refueling at Elefsina Airport. *No succession in aerial coordination.*

17:45 S64N–154 helicopter, the single aircraft operating on the fire, informs the Leadership (Operational and Political) of the Fire Department and the Director of ESKE again, that the fire has entered the pine forest in Pentelis, which is not accessible on the ground by any vehicles and there is very dense smoke.

A fire department aircraft CL–415 (2054) takes off for Ntaou Pentelis.

17:46 A fire department heavy duty aircraft S64N–194 takes off for Ntaou Pentelis.

17:56 A fire department aircraft CL–415 (2042) takes off for Ntaou Pentelis.

17:57 The Eastern Attica Fire Department Commander (codename ARIS EASTERN ATTICA) arrives at N. Voutzas.

18:00 Fire department aircraft S64N–194 is diverted to MOTOR OIL by order of ESKE and the Chief of the Fire Department.

Evacuation of the Lyrio Foundation.

There are nine fire engines on location in Ntaou Pentelis where the fire began, two fire engines in Kallithechnoupoli where the first fire front spread, only one in N.Voutsas where the second fire front has spread and one in Agios Andreas.

A single helicopter operates on the fire.

18:05 The fire front enters N. Voutzas, on Riga Ferraiou Street no. 48.

Due to the ionization of the (DEDDIE) main electricity substation at that location, there is a drop in the main voltage supply in the area.

The Police block off Marathonos Av. at two points: a) first entrance of N. Voutzas b) second entrance of N. Voutzas.

18:10 A Chinook helicopter from Pachi Airport departs for Ntaou Pendelis.

18:12 Orders for a second S–64 for the Ntaou Pentelis fire to take off.

18:13 The fire front arrives at Lyrio Foundation.

18:15 Orders for CL–415 (2056) to divert from the Kinetta fire to Ntaou Pendelis.

18:20 The Municipal Representative of the Civil Protection Agency arrives in N. Voutsas.

The Police close off Marathonos Av. at three points: a) crossing with Fleming St., b) crossing with Aeroporias St. in Agios Andreas and c) 102 St. in Agia Marina.

18:21 The Traffic Monitoring and Control Centre informs that the fire has crossed the Marathonos Av.

Call for the closing off of coastal roads.

18:28 The second heavy duty fire department aircraft S–64 takes off, from Elefsina Military Airport for Ntaou Pentelis.

Thirteen firefighting water carrier vehicles of the Fire Department are operating on the fire. Nine are located at Ntaou Pentelis, two in Kallitechnoupoli, one on Marathonos Av. by the entrance to Kallitechnoupoli and only one in N. Voutzas.

18:30 The fire front crosses the Marathonos Av. on the height of Isminis St. (N. Voutza), Kyanis Aktis(Mati), and on Tritonos St. (Mati)—at the location of Taverna O VASSILIS—and spreads into Mati.

The Police close off the roundabout in Kokkino Limanaki and the roundabout in Agios Andreas.

Helicopter Floga 10 takes off from Elefsina Airport.

18:39 The first heavy type S–64 154 helicopter, which was operating at Ntaou Pentelis from 17:10, lands at Tatoi Airport.

18:44 The fire spreads to Kokkino Limanaki, crossing the Marathonos Av. at the location of the Sanctuary Church of the Ascension in Rafina, very close to the main Rafina intersection that leads to the Port.

18:45 The fire reaches the coastline in Mati.

The fire threatens the RAMADA ATTICA RIVIERA Hotel, on Poseidonos Av. (the coastal road) at No. 55–57, where hundreds of people have found refuge.

17:00

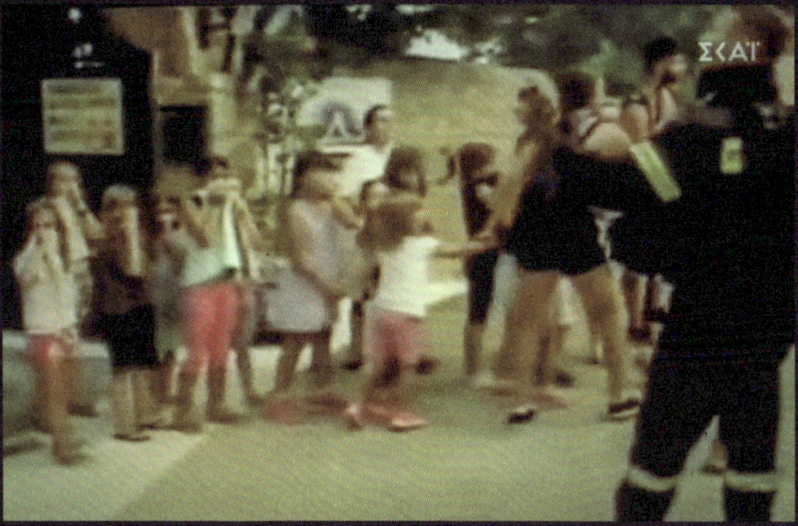

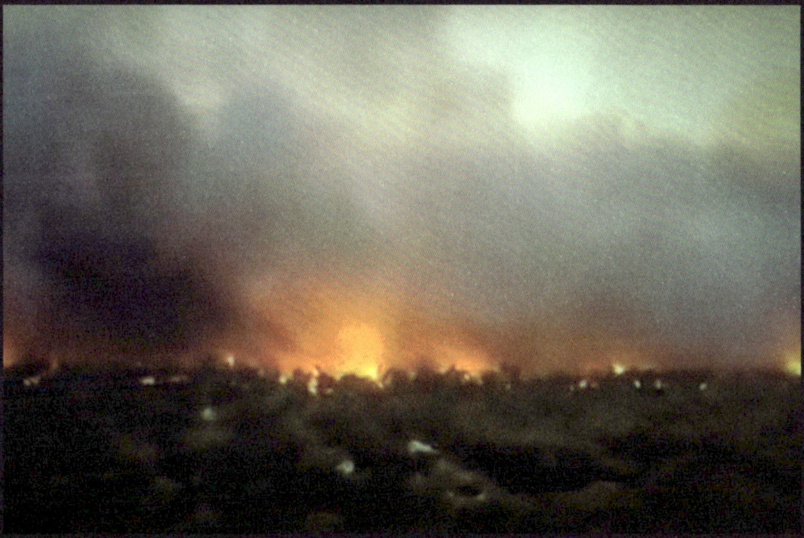

17:30

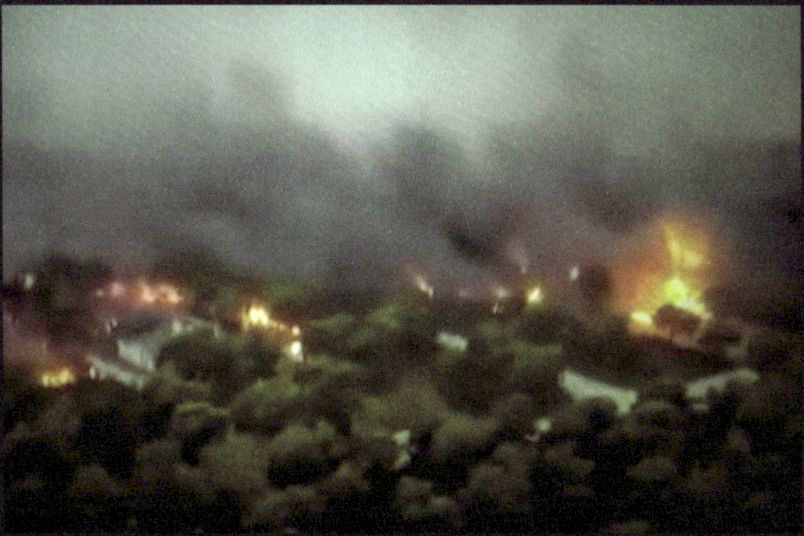

17:31

18:45

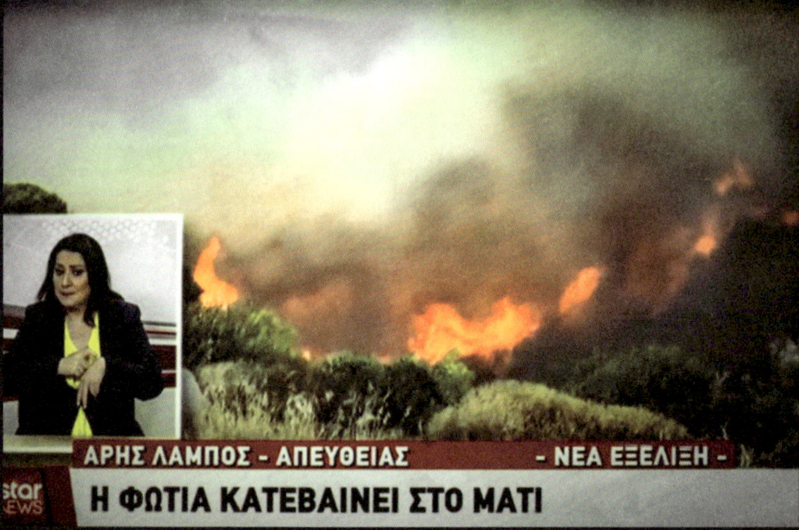

19:11

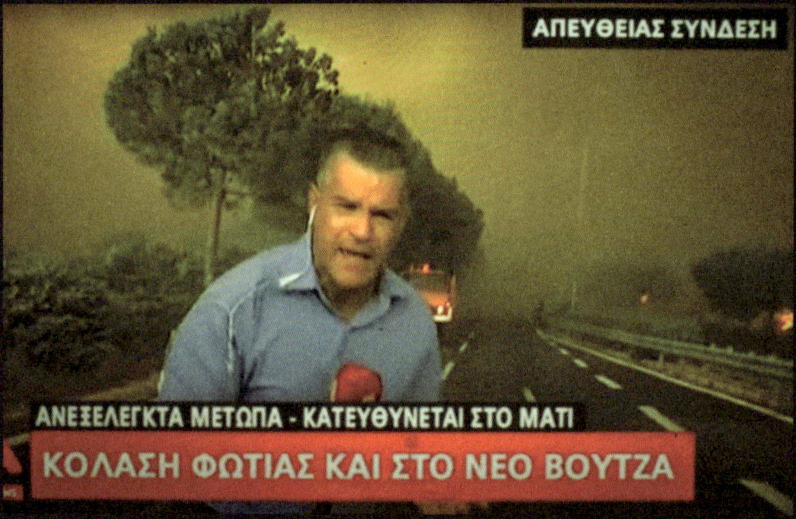

19:17

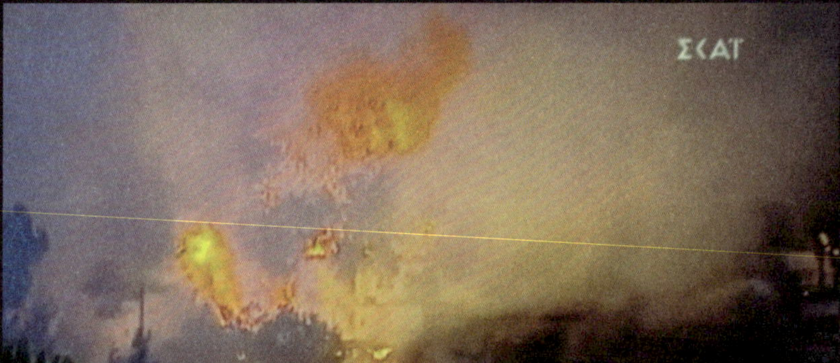

21.36

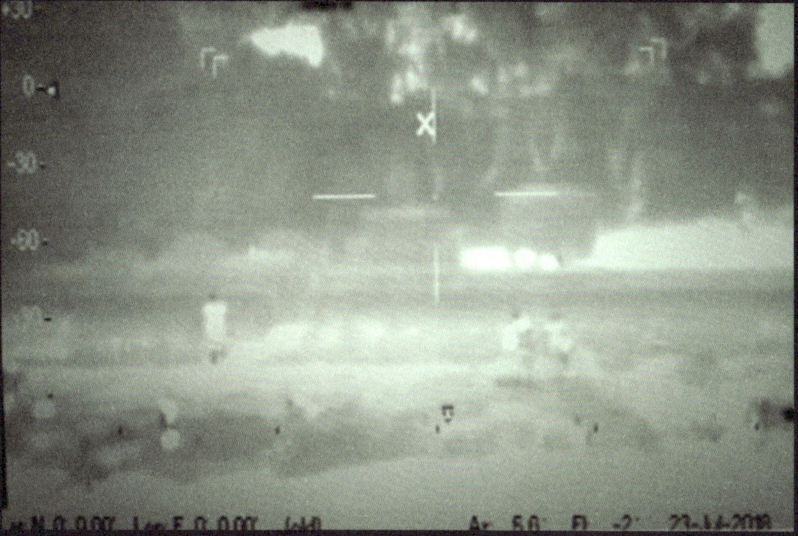

21.37

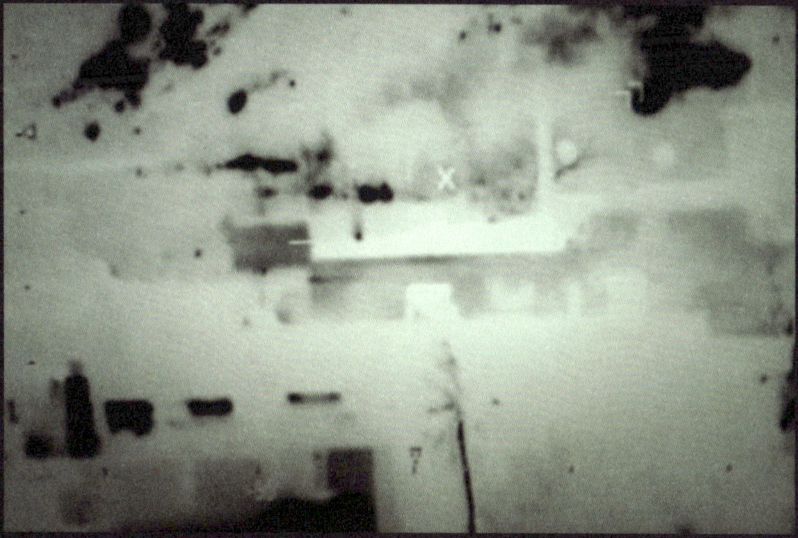

23.19

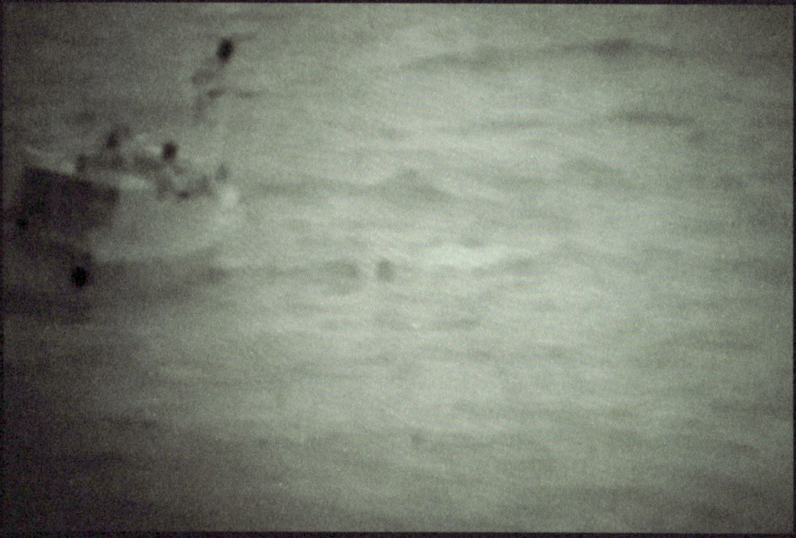

18:50 The Coast Guard is informed by a citizen that a large number of people are stranded on the various beaches, rocks and the sea due to the fire.

19:00 There are eighteen fire fighting water carrier vehicles combating the fire.

19:11 Helicopter A/F CL-415 (2054) informs the Leadership of the Fire Department and the Director of ESKE that there is a very large single front with very large flames 10–20 metres high, and the fire is moving toward the sea. The wind is southeastern and is moving the fire towards houses.

19:12 The Operations Officer of the Forest Fire Coordination Centre informs the Commander of ESKE that she will stop logging in aircraft traffic because she has lost the sequence, and the log will be filled in the end.

19:17 The fire front spreads further to Kokkino Limanaki crossing Marathonos Av. at the location of Agia Barbara.

19:29 Order for another S–64 to take off from the Elefsina Military Airport to Ntaou Pentelis.

19:42 A heavy-duty S–64 helicopter takes off from the Elefsina Military Airport for Ntaou Pentelis.

19:51 The Fire Department reports and confirms the first deceased in Dimokratias Av. 107 in Kokkino Limanaki.

19:52 A civilian (who is stranded on the beach in Argyra Akti/Kokkino Limanaki) informs ESKE that a teenage girl has fallen from a 20m cliff, is lying on the rocks and is in need of urgent care. Location of incident: Dimokratias Av. 162.

19:53 ESKE orders all ambulances to be directed to Mati.

20:06 The Director of ESKE requests the support of the Coast Guard for the rescue of people from the various beaches.

20:30 The Eastern Civil Protection Coordinating Body (SOPP) of Eastern Attica meets. The meeting ends at 21:00. Most representatives are absent, because they are operating on the fire.

20:36 Conversation of Chief of KEPP (Civil Protection Operation Centre) with Chief of 199 (Fire Department services) in which it is clear that the ambulances are afraid of entering the affected area. Order has been given to be escorted by the Fire Department.

20:36 Chief of the Special Disaster Response Unit (EMAK) arrives in Mati.

21:40 An employee of KEPP on a call to his mother says that there are deceased people in the areas affected but that they are not informing the public.

22:06 ESKE is informed that there are two deceased children within a house on the second entrance of the N.Voutsas settlement.

22:13 ESKE is informed that there are four deceased people in the RAMADA Hotel, and one person on Dimokratias Av. 109a.

22:14 KEPP is informed of deceased and severely burned people in the RAMADA Hotel.

22:41 ESKE is informed that there is one deceased person on the intersection of Nymfon St. and Poseidonos Av. in Mati.

22:45 ESKE is informed that there is one deceased person on the intersection of Chrisostomou Smyrnis and Thetidos St.

23:19 ESKE is informed that there are two deceased foreign nationals in the port of Rafina, one deceased woman on 20 Triados St., one deceased person on Dimokratias Av., four deceased people, two adults and two children, on Ag. Ioannou St. in N.Voutsas.

23:21 ESKE is informed of four deceased people in the intersection of Tritonos St. and Poseidonos Av., and one deceased person on the intersection of Perikleous St. and Poseidonos Av. in Mati.

23:31 ESKE is informed of one deceased person on the intersection of Karpenisiou St. and Marathonos Av. in N.Voutsas.

23:37 The Prime Minister Mr. Alexis Tsipras arrives at ESKE to be briefed. The briefing is broadcasted live on TV. In the briefing the Ntaou Pendelis/Mati fire is not given priority, the scale of the disaster and the number of deceased or injured is not mentioned.

23:57 A conversation is recorded in which it is mentioned that there are more than thirty deceased people.

Evacuations from the various beaches via the sea proceed, as well as search and rescue missions on the ground, in the sea and from the air.

24 July 2018 Search and rescue missions continue throughout the night. The Army is mobilized, along with the Red cross and all other emergency services.

16:28 The Minister of Defence, Panos Kammenos, arrives to the area in the afternoon. In the interview he gives to the BBC from the cove of Argyra Akti—where a number of people died—he directly accuses the residents of closing off access to the sea—'they have taken over the shore (by building) without any regard for the laws.'

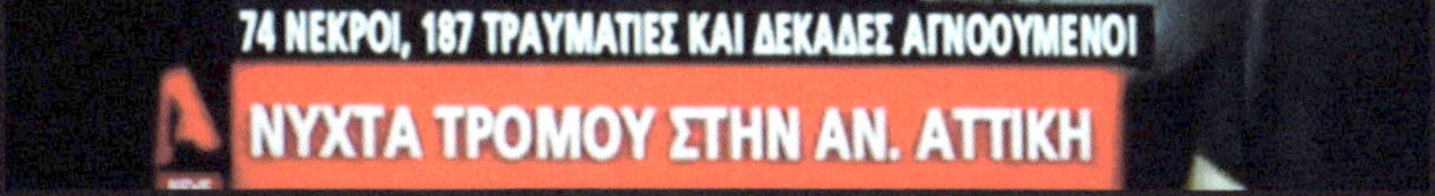

23.37

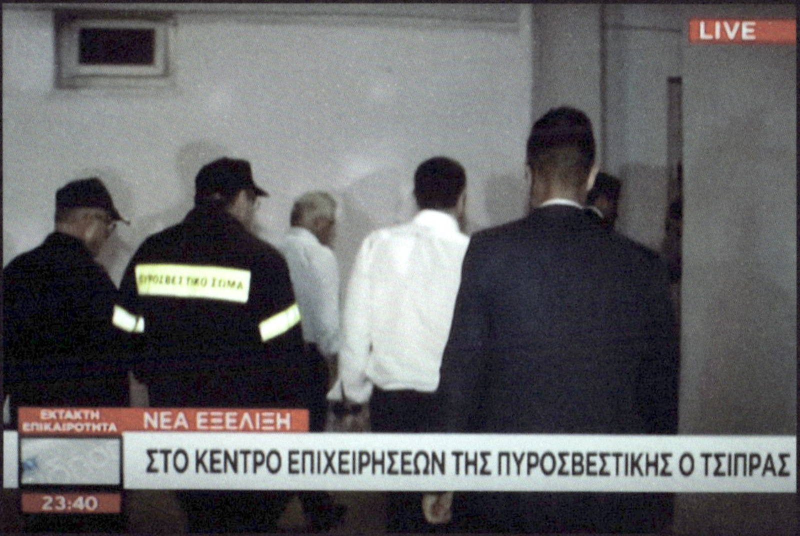

24/07/18
16:28

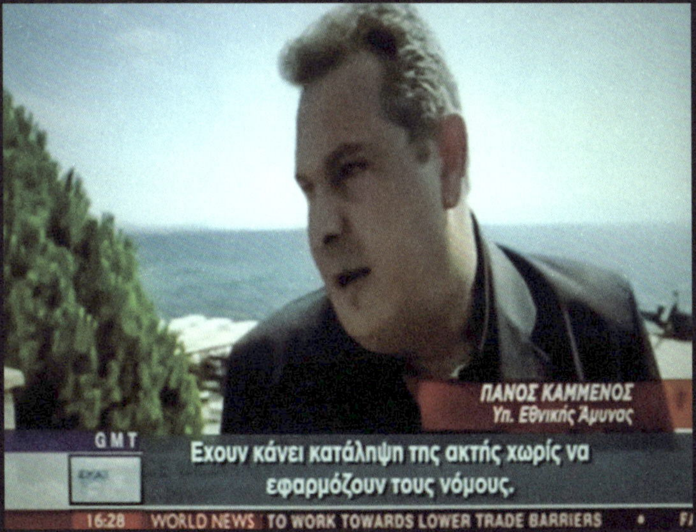

Aircraft Log

Normal estimated prep time for all aircrafts after order for take off: 20 minutes
Normal estimated prep time for Eriscon aircraft after order for take off: 15 minutes

Flight time from Elefsina to Ntaou Pendelis: 20 minutes
Flight time to gather water after reaching the location of Ntaou Pendelis: 10 minutes

Chinook Helicopters cannot operate in inhabited areas.
Eriscon S64 Helicopters can operate in inhabited areas
and make precision targeted water drops.

Aircraft (helicopters and planes) that do not operate on the fire with ignition location location in Ntaou Pentelis.

Aircraft	Time	Stationed airport	Status
BK117–FLOGA 2	—	Eleftherios Venizelos	Does not receive orders to operate.
FLOGA 3	—	Eleftherios Venizelos	Does not receive orders to operate.
KA 32	15:55 17:45	Chios—Elefsina	Departs Chios Island. Lands at Elefsina airport. Does not receive orders to operate.
PETSETEL 2806 Airforce	—	Tatoi	Does not receive orders to operate.
PETSETEL 28	— 17:30	Lamia	Ordered to Sernikaki Amfissa fire. Not used. Returns to airport.
PETSETEL 30	— 17:30	Lamia	Ordered to Sernikaki Amfissa fire. Not used. Returns to airport.
CHINOOK 923	—	Pachi Megara	Does not receive orders to operate.
CL215 1039	16:04 17:40 18:12 18:24 18:28	Elefsina	Lands in Elefsina Airport. Declares Fault 'out of order flaps.' Error restored. Ordered to Ntaou Pendelis fire. Cancels take off due to Fault 'out of order.'
CL215		Samos	Available, cannot take off due to winds.
CL215	—	Andravida	Available, cannot take off due to winds.
CL 215 1110	17:45 18:45	Andravida	Departs for the Kineta fire. Diverted to Isthmia fire.
TESNA Coast Guard	—	—	Available, cannot take off due to winds.

Aircraft that operate on the fire with ignition location in Ntaou Pentelis.

Aircraft	Time	Stationed airport	Status	Water drops
FLOGA 1 Coordinating Helicopter No Firefighting Capability	16:59 17:17 17:21–17:25 17:31	Elefsina	Diverted to Ntaou Pendelis fire. Arrives in Ntaou Pendelis. Assumes coordination responsibilities. 3 Communications with Operational Head on the ground. Departs from Ntaou Pendelis and lands in Elefsina.	—
S64 154	16:50 17:10 17:30 18:20 18:39	Tatoi	Diverted from Kineta to Ntaou Pendelis fire. Only aircraft to operate on Ntaou Pendelis fire. Informs that fire is outside homes. Informs that fire is moving east, and will threaten homes. Informs that fire is within an inhabited area. Departs from Ntaou Pendelis fire. Lands at Tatoi Airport.	25
S64 N194	16:23 16:46 16:50 17:46 18:01	Elefsina	Lands at Elefsina Airport. Ordered to Ntaou Pendelis fire. Cannot take off due to winds. Departs for Ntaou Pendelis fire. Diverted from Ntaou Pendelis to Motor Oil.	1
CL 415 2054	15:48 17:20 17:45 20:01	Tanagra	Lands at Tanagra Airport. Ordered to Ntaou Pendelis fire. Departs for Ntaou Pendelis fire. Reports a fault and returns to Tanagra airport.	8
CL415 2052	15:51 17:20 17:56 21:15	Tanagra	Lands at Tanagra Airport. Ordered to Ntaou Pendelis fire. Departs for Ntaou Pendelis with ETA 18:11. Returns to Tanagra Airport.	11
CL415 2042	15:50 17:20 17:56 18:07 18:27	Tanagra	Lands at Tanagra Airport. Ordered to Ntaou Pendelis fire. Departs for Ntaou Pendelis fire. Arrives at Ntaou Pendelis. Returns with Fault to Tanagra Airport.	2
SINOUK 922	16:12 17:01 18:10 20:25	Pachi Megara	Lands at Pachi Airport due to low visibility. Ordered to Ntaou Pendelis fire. Departs for Ntaou with ETA 18:30. Returns to Pachi Airport.	14
CL415 2056	17:05 18:55 19:05	—	Operates on the Megara fire. Diverted from Kineta fire to for Ntaou Pendelis fire. Reports Fault and returns to airport.	3
S64 189	16:46 17:39 18:28	Elefsina	Reports that fire in Isthmia is on low bushes. Lands at Elefsina Airport. Departs for Ntaou Pendelis fire.	36
S64 218	17:00 17:38 19:42 21:15	Elefsina	At Elefsina airport. Cannot take off due to wind. Ordered to Ntaou Pendelis fire. Returns to Elefsina airport.	15
FLOGA10	17:28 18:46	Pachi	Lands in Pachi airport. Departs for Ntaou Pendelis fire. Reports that it performed 37 water drops in falsified document.	37?

A. Also, according to article 2 par. 1 and 2, of the same law as above, Disaster means any rapidly or slowly developing natural phenomenon or technological event in land, sea and air space, which causes widespread adverse effects on humans, as well as on the man-made or natural environment. The intensity of the disaster is determined by the size of the losses or damages concerning the lives, health and property of citizens, goods, productive sources and infrastructures while Risk means the possibility of a natural phenomenon or technological event or other disasters combined with the intensity of the disasters, which may be caused to citizens, goods, wealth-producing sources and infrastructure of an area.

[...]

B. Furthemore, on page 73, of the above-mentioned Planning Document in paragraph 17 'SETTLEMENTS OR AREAS OF ARBITRARILY BUILT URBAN AREAS IN WILD LAND-URBAN INTERFACE ZONES', the specific areas of arbitrary construction in Attiki are recorded based on Urban Planning data. It is of essence that these areas where the fatal fire occurred ,such as Neos Voutsas and Matiof the Municipality of Marathon, are not mentioned.

A. Law 3013/2002 Article 2
Definition of what constitutes a Disaster.

B. From the Forest Fire Response Plan, prepared by the Regional Fire Administration (PE.PY.D.) of Attica, and approved on 25-05-2018b by the Operational Deputy Chief of the Fire Department, Vasilios Matthaiopoulos, Lieutenant General of the Fire Department.

Aftermath, Cover up and Legal case

Mr. Dimitris Liotsios was assigned as the Forensic Expert by the Primary Prosecutor's office of Athens and Piraeus for the Criminal Case of the Fire that erupted on the 23 July 2018 and approximate time of 16:40 in the area of Ntaou Pendelis and spread to areas of Kalltechnoupoli, N. Voutzas, Mati and Kokkino Limanaki according to the assigning documents No 171.381 of the 26-07-2018 and 171.905 of the 03-08-2018.

According to his testimony 'In the context of the meddling with my work and before submitting my report, I received a phone call from the new Chief of the Fire Department, Mathaiopoulos Vasilios of Konstantinou, on 20-09-2018 at 23:41 [...]. In this phone call, Mr. Chief, in a strong tone, ordered me to go to his office the next day because we had to talk, as he told me, about something serious. Indeed, the next day, at around 11:00, I presented myself at the office maintained by the Chief of the Fire Department at the E.S.K.E.[...].'

In the evidence that he submitted, that were also released in the *Kathimerini* newspaper on 19-07-2020 (youtube.com/watch?v=npNqo3b28lc), Mr. Liotsios was threatened, and intimidated into omitting information from his report and ordered to stop his further investigation. Mr. Liotsios, refused to succumb to the pressure and threats, and proceeded to further his investigation and depositions in front of the later court with all the evidence he had gathered.

Evidence that was instrumental in proving the events of the day.

First trial started on the 22 October 2022.

The charges focused on omissions and negligence on the part of the competent authorities and public officials, who were considered to have failed to react promptly and effectively to the crisis. Among the defendants were high-ranking commanders of the Fire Department, as well as various figures from the local authorities. In total, twenty-one people faced misdemeanor charges including manslaughter and negligence causing bodily harm. There were at least three attempts to elevate the charges to felonies, but were rejected.

An investigation into the role and responsibility of the Coast Guard, was ordered after the request of a number of family members of those victims that drowned at sea. The case was investigated by the Navy Court Martial, and the Coast Guard acquitted of any wrongdoing. A second and third investigator were ordered, and again the Coast Guard was acquitted. The Civilian Courts have no jurisdiction to trial any member of the Coast Guard.

An investigation or inquiry in the role and responsibility of the Police Department was never ordered by anyone.

The trial was concluded in 2024 with the issuance of the first verdicts. Six of the twenty-one defendants were found guilty of manslaughter and negligent bodily harm, while the rest were acquitted.

The following defendants were convicted:

1. Sotiris Terzoudis, then Chief of the Fire Department.
2. Vassilis Matthaiopoulos, then Deputy Chief of the Fire Department.
3. Ioannis Fostieris, Commander of ESKE.
4. Nikolaos Panagiotopoulos, Commander of the Athens Fire Department.
5. Charalambos Chionis, Commander of the East Attica Fire Department.
6. Konstantinos Angelopoulos, resident of Ntaou Pendelis that started the fire.

The sentences given ranged from three to 111 years. However, the actual sentences were significantly shorter and were eventually converted into redeemable ones, with the cost amounting to ten euros per day of imprisonment.

The Athens Prosecutor's Office of Appeals, following the issuance of the decision, proceeded to its appeal. This appeal covers both the sentences imposed on the convicted and the acquittals that were decided.

The second trial at the Three Judge Court of Appeals for Misdemeanors started on 08 July 2014.

All twenty-one defendants are currently being re-trialed.

The Public Prosecutor in her closing argument named those that should be found guiltly and those that should be acquitted according to evidence. Her suggestion, which go against the previous verdict is that members of the Political Leadership, Municipal, Regional and Central should be, according to evidence, found guilty. In total she suggested the conviction of twelve out of the twenty-one defendants, including those previously convicted.

The verdict is expected in the summer of 2025.

The Result is expected to be again contested and a retrial to start shortly after at the Supreme Court.

Unless the trial at the Supreme Court is concluded by the summer of 2026, the case's stature of limitations will run out and the case might be dismissed overall.

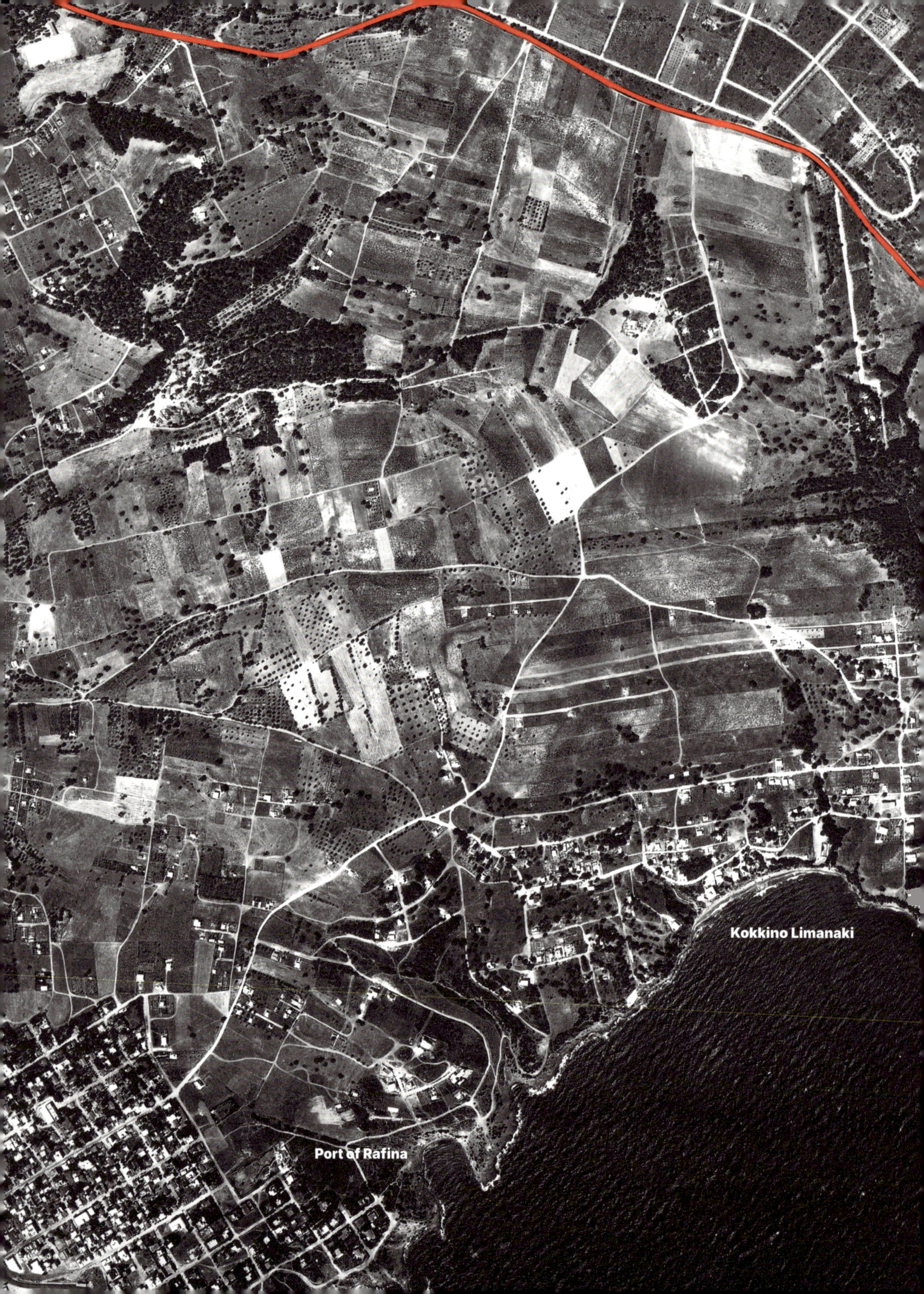
Kokkino Limanaki
Port of Rafina

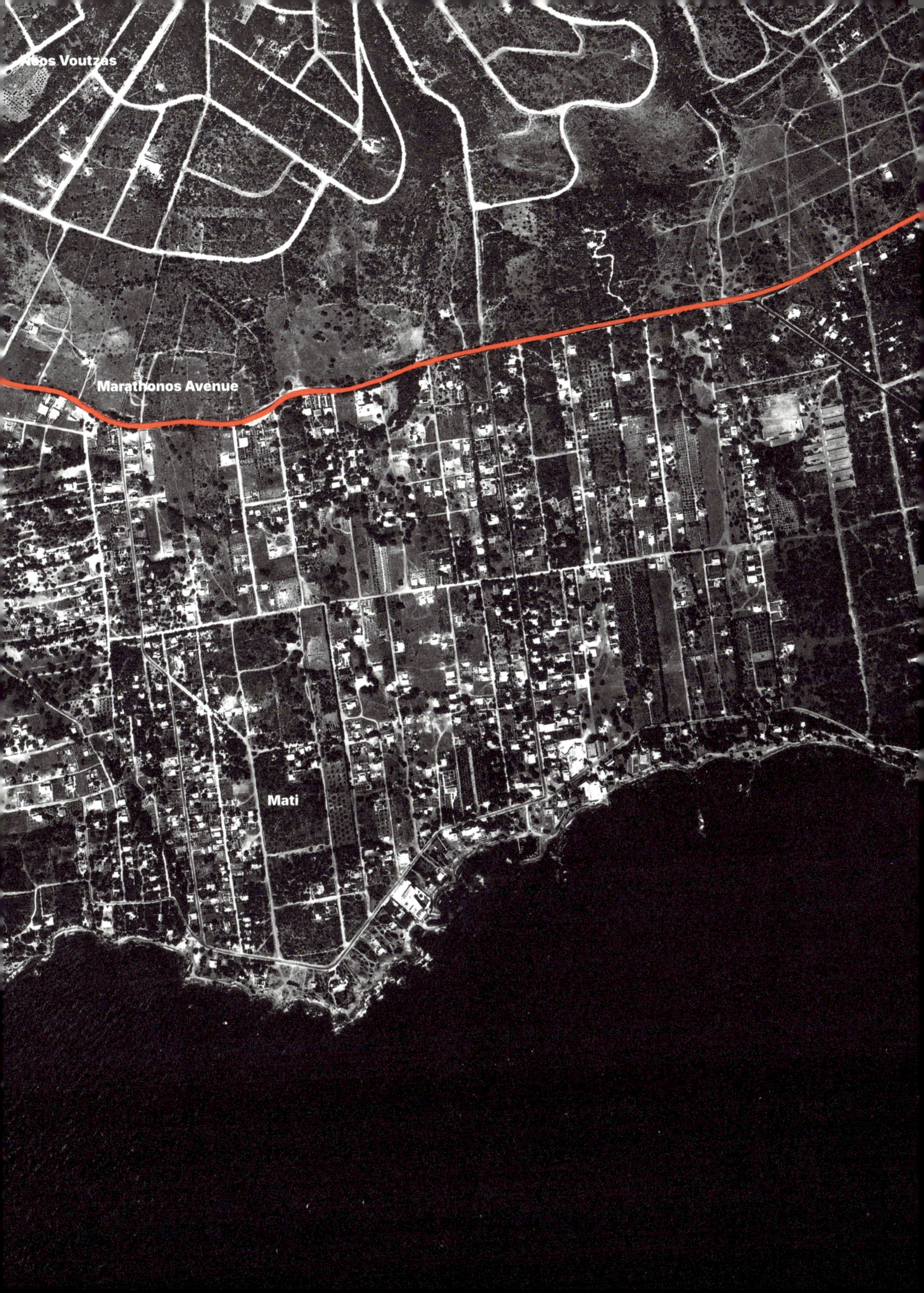

Neos Voutzas
Marathonos Avenue
Mati

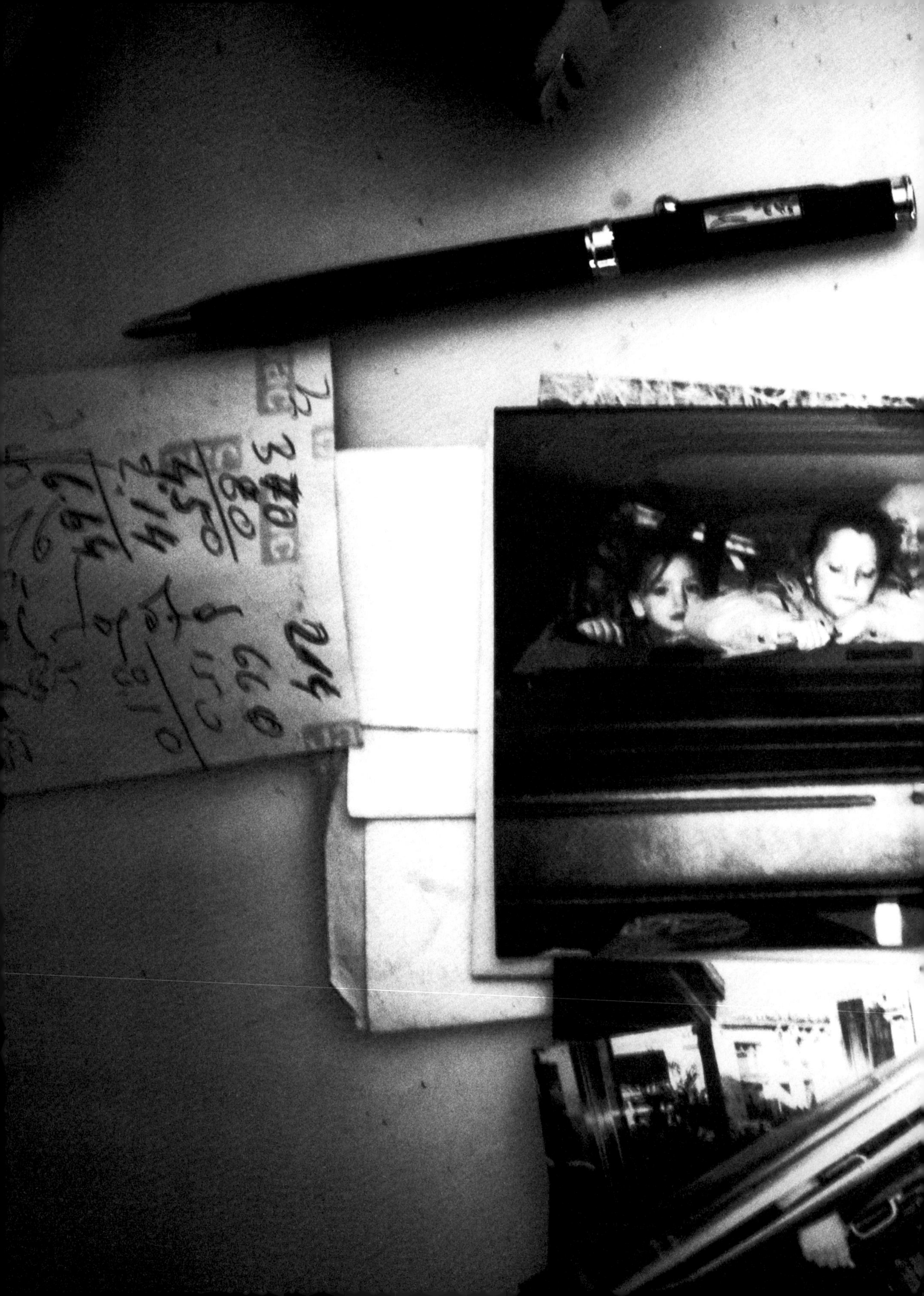

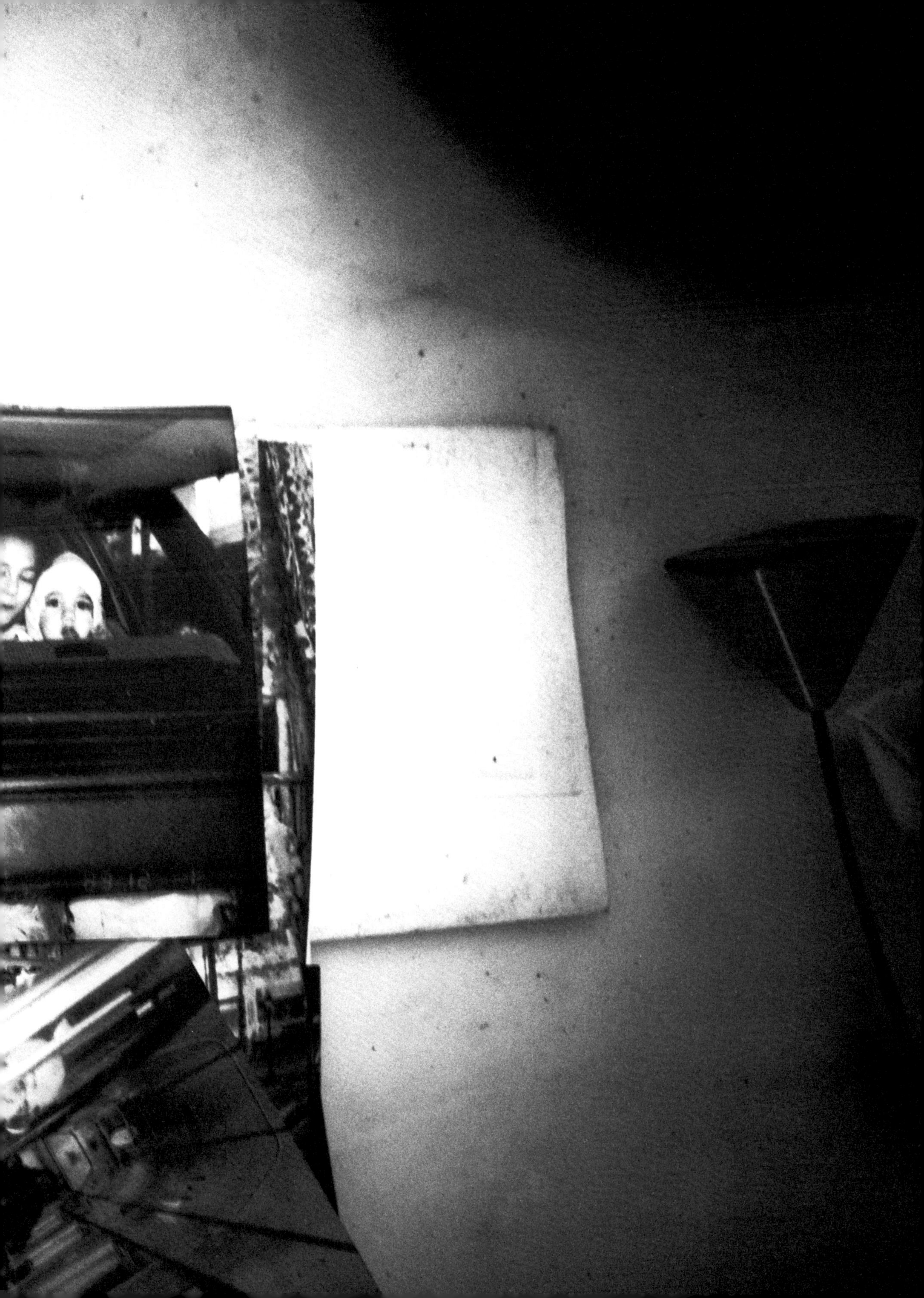

Unofficial list of identified victims 104*

Αβραμίδου Αικατερίνη
Avramidi Aikaterini
From: Greece
Gender: Female
Age: 47
Cause of Death: Extensive 5th degree burns

Αθανασοπούλου Ζωή
Athanasopoulou Zoi
From: Greece
Gender: Female
Age: 86
Cause of Death: Unknown

Αλεξόπουλος Δημήτρης-Αριστείδης
Alexopoulos Dimitris-Aristeidis
From: Greece
Gender: Male
Age: 13
Cause of Death: Unknown

Ανδρουλιδάκη Αιμιλία
Androulidaki Aimilia
From: Greece
Gender: Female
Age: 73
Cause of Death: Drowning

Βονικόπουλος Ευθύμιος
Vonikopoulos Eythimios
From: Greece
Gender: Male
Age: 55
Cause of Death: Drowning

Bonte Dirk
From: Belgium
Gender: Male
Age: 64
Cause of Death: Drowning

Γαζεπίδου Γαπέζη Παναγιώτα
Gapenidou Gapezi Panagiota
From: Greece
Gender: Female
Age: 54
Cause of Death: Extensive burns

Γαλάνη Μαρία
Galani Maria
From: Greece
Gender: Female
Age: 41
Cause of Death: Unknown

Γιαννακόδημος Βασίλειος
Giannakodimos Vasileios
From: Greece
Gender: Male
Age: 76
Cause of Death: Multiple organ failure due to extensive burns

Γιαννακοπούλου Σοφία
Giannakopoulou Sofia
From: Greece
Gender: Female
Age: 86
Cause of Death: Unknown

Γιαννοπούλου Αγγελική
Giannakopoulou Aggellki
From: Greece
Gender: Female
Age: 88
Cause of Death: Carbonisation

Δεσύπρη Θέκλα
Desypri Thekla
From: Greece
Gender: Female
Age: 61
Cause of Death: Carbonisation

Δημητρίου αβάπτιστο
Dimitriou
From: Greece
Gender: Unknown
Age: 6 months
Cause of Death: Thermal burns

Διονυσιώτη Μαργαρίτα
Dionisioti Margarita
From: Greece
Gender: Female
Age: 35
Cause of Death: Multiple organ failure due to extensive burns

Δόξας Αργύριος
Doxas Argyris
From: Greece
Gender: Male
Age: 60
Cause of Death: Carbonisation

Δόξας Παναγιώτης
Doxas Panagiotis
From: Greece
Gender: Male
Age: 20
Cause of Death: Carbonisation

Δρούβης Χαράλαμπος
Drouvis Charalambos
From: Greece
Gender: Male
Age: 73
Cause of Death: Extensive burns

Ζαφειρίου Ευστρατίος
Zafeiriou Eustratios
From: Greece
Gender: Male
Age: 85
Cause of Death: Extensive burns

Θάνος (Ντανος) Ιωάννης
Thanos (Ntanos) Ioannis
From: Greece
Gender: Male
Age: Uknown
Cause of Death: Unknown

Κάκαρης Σπύρος
Kakaris Spyros
From: Greece
Gender: Male
Age: 40
Cause of Death: Unknown

Καλαμπανίδα Μαρία
Kalampanida Maria
From: Greece
Gender: Female
Age: 85
Cause of Death: Carbonisation

Καλαμπανίδα Μαριάνθη
Kalmpanida Marianthi
From: Greece
Gender: Female
Age: Uknown
Cause of Death: Carbonisation

Καλεγγιαννάκος Νικόλαος
Kaleggianakos
From: Greece
Gender: Male
Age: 85
Cause of Death: Carbonisation

Καλεγγιαννάκος Ιωάννης
Kaleggianakos Ioannis
From: Greece
Gender: Male
Age: 51
Cause of Death: Thermal burns

Καλεγγιαννάκου Παρασκεύη
Kaleggiannakou Paraskeui
From: Greece
Gender: Female
Age: 74
Cause of Death: Extensive 5th degree burns

Κανδαλέπα Σμαράγδα
Kandalepa Smaragda
From: Greece
Gender: Female
Age: 95
Cause of Death: Extensive burns

Καραγκιαούρη Γαρυφαλία
Karagkiaouri Garyfalia
From: Greece
Gender: Female
Age: 66
Cause of Death: Unknown

Καραλή Αθηνά (Νέλλη)
Karali Athina (Nelly)
From: Greece
Gender: Female
Age: 48
Cause of Death: Extensive burns

Κατσαργύρης Βασίλειος
Katsargyris Vasileios
From: Greece
Gender: Male
Age: 70
Cause of Death: Carbonisation

Κατσουλάκης Μαρκος
Katsoulakis Markos
From: Greece
Gender: Male
Age: 53
Cause of Death: Unknown

Κατσουλάκης Πολύδωρος (Πάρης)
Katsoulakis Polydoros (Paris)
From: Greece
Gender: Male
Age: 20
Cause of Death: Thermal burns

Κέρτσος Κωνσταντίνος
Kertsos Konstantinos
From: Greece
Gender: Male
Age: 83
Cause of Death: Unknown

Κιντάνζη Μαργαρίτα
Kintanzi Margarita
From: Greece
Gender: Female
Age: 69
Cause of Death: Died in Hospital after 4 years of Hospitalization

Κλάδιος Ιάκωβος
Kladios Iakovos
From: Greece
Gender: Male
Age: 47
Cause of Death: Carbonisation

Κλάδιος Ιωάννης
Kladios Ioannis
From: Greece
Gender: Male
Age: 84
Cause of Death: Carbonisation

Κλάδιου Μαρία
Kladiou Maria
From: Greece
Gender: Female
Age: 81
Cause of Death: Carbonisation

Κοκκινίδης Γεώργιος
Kokkinidis Georgios
From: Greece
Gender: Male
Age: 8
Cause of Death: Carbonisation

Κοκκινίδης Δημήτριος
Kokkinidis Dimitrios
From: Greece
Gender: Male
Age: 5
Cause of Death: Carbonisation

Κοκκινίδης Παναγιώτης
Kokkinidis Panagiotis
From: Greece
Gender: Male
Age: 39
Cause of Death: Unknown

Κοκκινίδου Πιπίνα
Kokkinidou Pipina
From: Greece
Gender: Male
Age: 67
Cause of Death: Carbonisation

Κοκκώνης Επαμεινώνδας
Kokkonis Epameinondas
From: Greece
Gender: Male
Age: 75
Cause of Death: Unknown

Korzeniowska Beata-Teresa
From: Poland
Gender: Female
Age: 38
Cause of Death: Drowning

Korzeniowsk Kacper-Nikolaj
Korzeniowsk Kacper
From: Poland
Gender: Male
Age: 9
Cause of Death: Drowning

Κοσσόρας Νικόλαος
Kossoras Nicholaos
From: Greece
Gender: Male
Age: 68
Cause of Death: Unknown

Κούκλα-Πολυχρονίδου Μαρία
Koukla-Polychronidou Maria
From: Greece
Gender: Female
Age: 71
Cause of Death: Carbonisation

Κούκλας Χρήστος
Kouklas Christos
From: Greece
Gender: Male
Age: 77
Cause of Death: Carbonisation

Κωνσταντάκη Μαριάνθη
Konstantaki Marianthi
From: Greece
Gender: Female
Age: 85
Cause of Death: Unknown

Κωστοπούλου Βασιλεία
Konstopoulou Vasileia
From: Greece
Gender: Female
Age: 78
Cause of Death: Extensive burns

Λαμπρίδης Μιχαήλ
Lampridis Michail
From: Greece
Gender: Male
Age: 73
Cause of Death: Unknown

Λιναρδάτος Στέλιος
Linardos Stelios
From: Greece
Gender: Male
Age: Unknown
Cause of Death: Unknown

Μάσχα Αγγελική
Masxa Aggeliki
From: Greece
Gender: Female
Age: 85
Cause of Death: Unknown

Μάσχας Άγγελος
Masxas Aggelos
From: Greece
Gender: Male
Age: 88
Cause of Death: Unknown

Μαυρίδου Βασιλεία
Mauridou Vasileia
From: Greece
Gender: Female
Age: 59
Cause of Death: Thermal burns

Μίχας Βίκτωρας
Mixas Victoras
From: Greece
Gender: Male
Age: 25
Cause of Death: Drowning

Μόσχου Μαρία
Mosxou Maria
From: Greece
Gender: Female
Age: 56
Cause of Death: Multiple organ failure due to extensive burns

Μωραϊτη Ασπασία
Moraiti Aspasia
From: Greece
Gender: Female
Age: 91
Cause of Death: Unknown

Νιτσοτόλη Κατερίνα (Κατίνα)
Nitsotoli Katerina (Katina)
From: Greece
Gender: Female
Age: 66
Cause of Death: Carbonisation

Ντούτσα-Γαβάκη Νικολέτα
Ntoutsa-Gavaki Nikoleta
From: Greece
Gender: Female
Age: 46
Cause of Death: Extensive burns

O'Calaghan Westropp Brian George
From: Ireland
Gender: Male
Age: 47
Cause of Death: Extensive Burns

Ορφανού Κωστούλα
Orfanou Kostoula
From: Greece
Gender: Female
Age: 82
Cause of Death: Thermal burns

Παγωμένου Μαρία
Pagomenou Maria
From: Greece
Gender: Female
Age: 73
Cause of Death: Carbonisation

Παλιούρα Βασιλική
Palioura Vasiliki
From: Greece
Gender: Female
Age: 73
Cause of Death: Drowning

Παντελιδάκης Στυλιανός-Ιωάννης
Pantelidakis Stylianos-Ioannis
From: Greece
Gender: Male
Age: 71
Cause of Death: Carbonisation

Παπαιωάννου Ζόγκα Floreta Φλώρα
Papaioanna Zoga Floreta Flora
From: Greece
Gender: Female
Age: 57
Cause of Death: Extensive burns and cerebral ischaemia

Παπαιωάννου Ιωάννης
Papaioannou Ioannis
From: Greece
Gender: Male
Age: 61
Cause of Death: Unknown

Παπαιωάννου Παναγιώτης
Papaioannou Panagiotis
From: Greece
Gender: Male
Age: 62
Cause of Death: Extensive burns

Ιερέας Παπαποστόλου Σπυρίδων
Priest Papapostolou Spyridon
From: Greece
Gender: Male
Age: 83
Cause of Death: Drowning

Πατελάρου Μοίρα
Pantelarou Moira
From: Greece
Gender: Female
Age: 87
Cause of Death: Carbonisation

Πεταλά Ευαγγελία
Petala Euaggelia
From: Greece
Gender: Female
Age: 71
Cause of Death: Thermal burns

Πεταλάς Διονύσιος
Petalas Dionysios
From: Greece
Gender: Male
Age: 89
Cause of Death: Unknown

Πλυμάκης Λεωνίδας
Plymakis Leonidas
From: Greece
Gender: Male
Age: 75
Cause of Death: Unknown

Πολίτου Ευγενία
Politou Eugenia
From: Greece
Gender: Female
Age: 77
Cause of Death: Carbonisation

Πολυπόρτης Μιχαήλ
Popyportis Michail
From: Greece
Gender: Male
Age: 66
Cause of Death: Unknown

Σαριέβα Ελένη
Sarieva Eleni
From: Greece
Gender: Female
Age: 56
Cause of Death: Carbonisation

Σιαπέρα Ευαγγελία
Saviera Euaggelia
From: Greece
Gender: Female
Age: 77
Cause of Death: Thermal burns

Σιαπκάρα Καλλιόπη
Saikara Kalliopi
From: Greece
Gender: Female
Age: 35
Cause of Death: Unknown

Σιδέρη Ελένη
Sideri Eleni
From: Greece
Gender: Female
Age: 93
Cause of Death: Drowning

Σκαραμαγκά Ζαμπέτα
Skaramagka Zampeta
From: Greece
Gender: Female
Age: 77
Cause of Death: Unknown

Σκαραμαγκάς Δημήτρης
Skaramagkas
From: Greece
Gender: Male
Age: 86
Cause of Death: Unknown

Σπανού Κασσιανή
Spanou Kassiani
From: Greece
Gender: Female
Age: 40
Cause of Death: Carbonisation

Σπηλιώτη Χρυσή
Soilioti Xrysi
From: Greece
Gender: Female
Age: 62
Cause of Death: Carbonisation

Σπυρίδης Σπύρος
Spyridis Spyros
From: Greece
Gender: Male
Age: 73
Cause of Death: Unknown

Σπυρόπουλου Θάλεια
Spyropoulou Thaleia
From: Greece
Gender: Female
Age: 68
Cause of Death: Multiple organ failure due to extensive burns

Τζούλιας Δημήτρης
Tzoulias Dimitris
From: Greece
Gender: Male
Age: 68
Cause of Death: Extensive burns and lung consolidation

Τουρναβίτης Δημήτρης
Tournavitis Dimitris
From: Greece
Gender: Male
Age: 54
Cause of Death: Thermal burns

Τσαούση Θεοδώρα
Tsaousi Theodora
From: Greece
Gender: Female
Age: 84
Cause of Death: Unknown

Τσάρμπος Παύλος
Tsarmpos Paulos
From: Greece
Gender: Male
Age: 60
Cause of Death: Unknown

Τσέκος Βίκτωρας
Tsekos Viktoras
From: Greece
Gender: Male
Age: 59
Cause of Death: Extensive burns

Φιλιπποπούλος Φίλιππος
Filippopoulos Filippos
From: Greece
Gender: Male
Age: 72
Cause of Death: Carbonisation

Φιλιπποπούλου Βασιλίκη
Filippopoulou Vasiliki
From: Greece
Gender: Female
Age: 9
Cause of Death: Unknown

Φιλιπποπούλου Σοφία
Filippopoulou Sofia
From: Greece
Gender: Female
Age: 9
Cause of Death: Unknown

Φιλιπποπούλου Σοφία
Filippopoulou Sofia
From: Greece
Gender: Female
Age: 61
Cause of Death: Unknown

Φύτρος Ανδρέας
Fytros Andreas
From: Greece
Gender: Male
Age: 11
Cause of Death: Thermal burns

Φύτρος Γρηγόρης
Fytros Grigoris
From: Greece
Gender: Male
Age: 54
Cause of Death: Thermal burns

Φύτρου Ευδοκια (Εβίτα)
Fytrou Eudokia (Evita)
From: Greece
Gender: Female
Age: 13
Cause of Death: Thermal burns and injuries sustained from jumping off a cliff

Χαμηλοθώρης Παναγιώτης
Xamilothoris Panagiotis
From: Greece
Gender: Male
Age: 29
Cause of Death: Extensive 5th degree burns

Χαρδαλούπα Ελισαβέτ
Xardaloupa Elisavet
From: Greece
Gender: Female
Age: 27
Cause of Death: Multiple organ failure due to severe thermal burns

Χαρδαλούπα Μαρία
Xardaloupa Maria
From: Unknown
Gender: Female
Age: Unknown
Cause of Death: Unknown

Χατζόπουλος Νικόλαος
Xatzopoulos Nicholaos
From: Greece
Gender: Male
Age: 84
Cause of Death: Extensive burns and soot inhalation

Χερουβείμ Βασιλική
Xerouveim Vasiliki
From: Greece
Gender: Female
Age: 53
Cause of Death: Carbonisation

Χερουβείμ Ευπραξία
Xerouveim Eypraxia
From: Greece
Gender: Female
Age: 74
Cause of Death: Carbonisation

Χερουβείμ Μελίνα-Ευπραξια (Εβελίνα)
Xerouveim Evelina
From: Greece
Gender: Female
Age: 5
Cause of Death: Carbonisation

Χερουβείμ Ηλιάνα-Μαρία (Μαρίλια)
Xerouveim Marilia
From: Greece
Gender: Female
Age: 5
Cause of Death: Carbonisation

Χριστοφίδου Στυλιανή (Στέλλα)
Xristofidou Styliani
From: Greece
Gender: Female
Age: 59
Cause of Death: Unknown

***7 April 2025**
An official list of fatalities (with or without data such as cause or locations of death) has never been comprised. Neither has a complete list of the hospitalized or injured.

It has long been known among the survivors, lawyers representing the case and some journalists that there is a mass grave in the Cemetery of Marathonas, in which various unidentified body parts of a number of different DNAs have been buried. There had been no further public information.

After persistent efforts to uncover the truth, it was revealed on the 7 April 2025 by the Association of Relatives of Victims and Burn Survivors that there are another 14–18 confirmed unidentified Victims. Their bodies never identified, or matched to any known DNA, or reported missing by anyone.

It is assumed that they were undocumented immigrant workers.

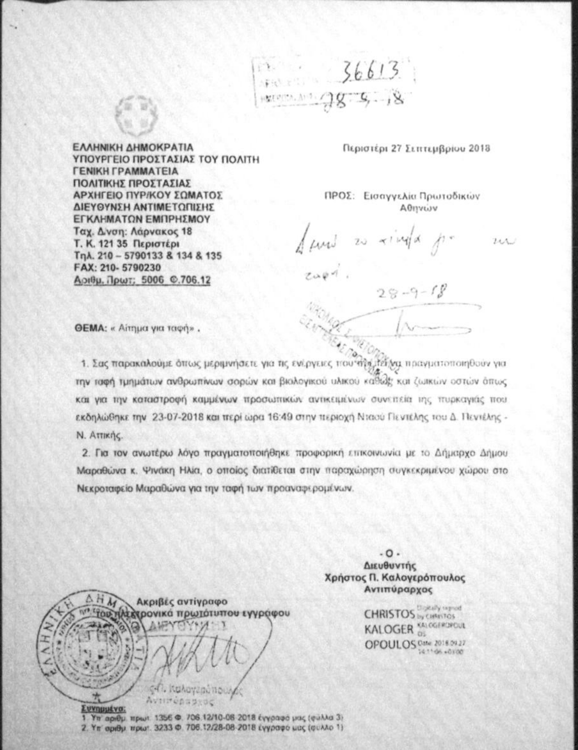

ΕΛΛΗΝΙΚΗ ΔΗΜΟΚΡΑΤΙΑ
ΥΠΟΥΡΓΕΙΟ ΠΡΟΣΤΑΣΙΑΣ ΤΟΥ ΠΟΛΙΤΗ
ΓΕΝΙΚΗ ΓΡΑΜΜΑΤΕΙΑ
ΠΟΛΙΤΙΚΗΣ ΠΡΟΣΤΑΣΙΑΣ
ΑΡΧΗΓΕΙΟ ΠΥΡ/ΚΟΥ ΣΩΜΑΤΟΣ
ΔΙΕΥΘΥΝΣΗ ΑΝΤΙΜΕΤΩΠΙΣΗΣ
ΕΓΚΛΗΜΑΤΩΝ ΕΜΠΡΗΣΜΟΥ
Ταχ. Δνση: Λάρνακος 18
Τ. Κ. 121 35 Περιστέρι
Τηλ. 210 – 5790133 & 134 & 135
FAX: 210- 5790230
Αριθμ. Πρωτ: 5006 Φ.706.12

Περιστέρι 27 Σεπτεμβρίου 2018

ΠΡΟΣ: Εισαγγελία Πρωτοδικών Αθηνών

ΘΕΜΑ: « Αίτημα για ταφή» .

1. Σας παρακαλούμε όπως μεριμνήσετε για τις ενέργειες που θα πρέπει να πραγματοποιηθούν για την ταφή τμημάτων ανθρωπίνων σορών και βιολογικού υλικού καθώς και ζωικών οστών όπως και για την καταστροφή καμμένων προσωπικών αντικειμένων συνεπεία της πυρκαγιάς που εκδηλώθηκε την 23-07-2018 και περί ώρα 16:49 στην περιοχή Νταού Πεντέλης του Δ. Πεντέλης - Ν. Αττικής.

2. Για τον ανωτέρω λόγο πραγματοποιήθηκε προφορική επικοινωνία με το Δήμαρχο Δήμου Μαραθώνα κ. Ψινάκη Ηλία, ο οποίος διατίθεται στην παραχώρηση συγκεκριμένου χώρου στο Νεκροταφείο Μαραθώνα για την ταφή των προαναφερομένων.

- Ο -
Διευθυντής
Χρήστος Π. Καλογερόπουλος
Αντιπύραρχος

CHRISTOS KALOGEROPOULOS

Ακριβές αντίγραφο
του ηλεκτρονικά πρωτότυπου εγγράφου

Συνημμένα:
1. Υπ' αριθμ. πρωτ. 1356 Φ. 706.12/10-08-2018 έγγραφό μας (φύλλα 3)
2. Υπ' αριθμ. πρωτ. 3233 Φ. 706.12/28-08-2018 έγγραφό μας (φύλλο 1)

Request from the Fire Department for the burial of a number of unidentified body parts and biological material, a result of the fire of the 23 July 2018.

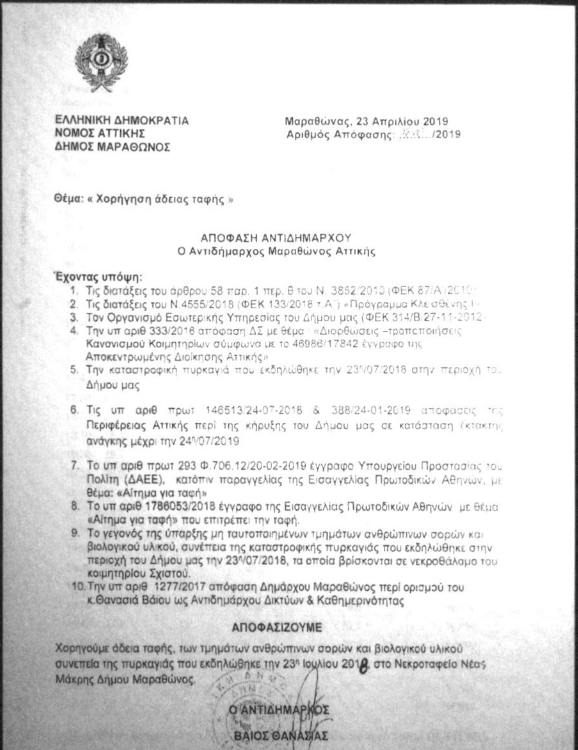

ΕΛΛΗΝΙΚΗ ΔΗΜΟΚΡΑΤΙΑ
ΝΟΜΟΣ ΑΤΤΙΚΗΣ
ΔΗΜΟΣ ΜΑΡΑΘΩΝΟΣ

Μαραθώνας, 23 Απριλίου 2019
Αριθμός Απόφασης: [illegible]/2019

Θέμα: « Χορήγηση άδειας ταφής »

ΑΠΟΦΑΣΗ ΑΝΤΙΔΗΜΑΡΧΟΥ
Ο Αντιδήμαρχος Μαραθώνος Αττικής

Έχοντας υπόψη:
1. Τις διατάξεις του άρθρου 58 παρ. 1 περ. θ του Ν. 3852/2010 (ΦΕΚ 87/Α/2010)
2. Τις διατάξεις του Ν.4555/2018 (ΦΕΚ 133/2018 τ.Α') «Πρόγραμμα Κλεισθένης Ι»
3. Τον Οργανισμό Εσωτερικής Υπηρεσίας του Δήμου μας (ΦΕΚ 314/Β/27-11-2012)
4. Την υπ αριθ 333/2016 απόφαση ΔΣ με θέμα : «Διορθώσεις –τροποποιήσεις Κανονισμού Κοιμητηρίων σύμφωνα με το 46986/17842 έγγραφο της Αποκεντρωμένης Διοίκησης Αττικής»
5. Την καταστροφική πυρκαγιά που εκδηλώθηκε την 23η/07/2018 στην περιοχή του Δήμου μας
6. Τις υπ αριθ πρωτ 146513/24-07-2018 & 388/24-01-2019 αποφάσεις της Περιφέρειας Αττικής περί της κήρυξης του Δήμου μας σε κατάσταση έκτακτης ανάγκης μέχρι την 24η/07/2019
7. Το υπ αριθ πρωτ 293 Φ.706.12/20-02-2019 έγγραφο Υπουργείου Προστασίας του Πολίτη (ΔΑΕΕ), κατόπιν παραγγελίας της Εισαγγελίας Πρωτοδικών Αθηνών, με θέμα: «Αίτημα για ταφή»
8. Το υπ αριθ 1786053/2018 έγγραφο της Εισαγγελίας Πρωτοδικών Αθηνών με θέμα «Αίτημα για ταφή» που επιτρέπει την ταφή.
9. Το γεγονός της ύπαρξης μη ταυτοποιημένων τμημάτων ανθρώπινων σορών και βιολογικού υλικού, συνέπεια της καταστροφικής πυρκαγιάς που εκδηλώθηκε στην περιοχή του Δήμου μας την 23η/07/2018, τα οποία βρίσκονται σε νεκροθάλαμο του κοιμητηρίου Σχιστού.
10. Την υπ αριθ 1277/2017 απόφαση Δημάρχου Μαραθώνος περί ορισμού του κ.Θανασιά Βάιου ως Αντιδημάρχου Δικτύων & Καθημερινότητας

ΑΠΟΦΑΣΙΖΟΥΜΕ

Χορηγούμε άδεια ταφής, των τμημάτων ανθρώπινων σορών και βιολογικού υλικού συνεπεία της πυρκαγιάς που εκδηλώθηκε την 23η Ιουλίου 2018, στο Νεκροταφείο Νέας Μάκρης Δήμου Μαραθώνος.

Ο ΑΝΤΙΔΗΜΑΡΧΟΣ

ΒΑΙΟΣ ΘΑΝΑΣΙΑΣ

Permission from the Municipality of Marathonas for the burial of such unidentified body parts and biological material.

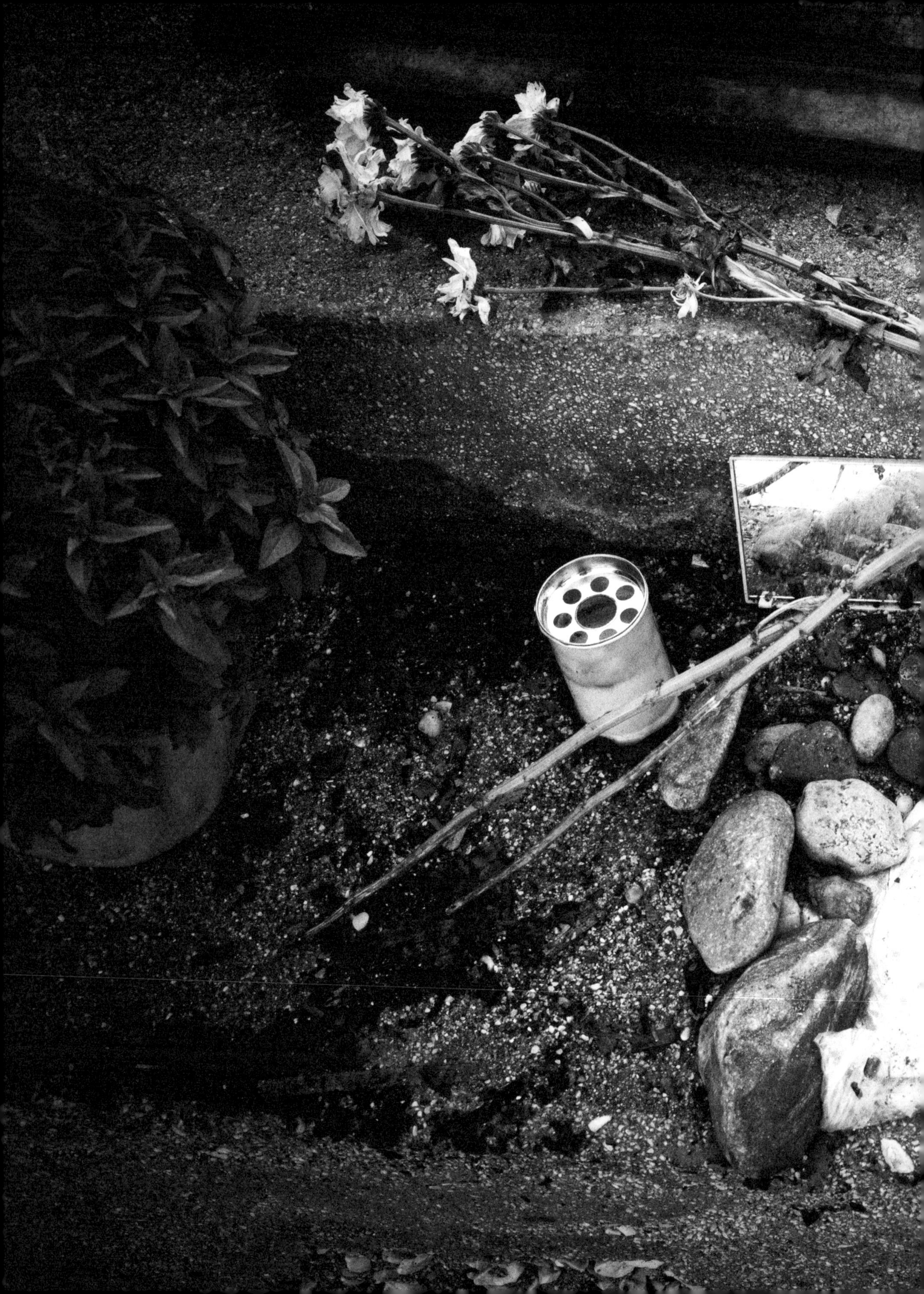

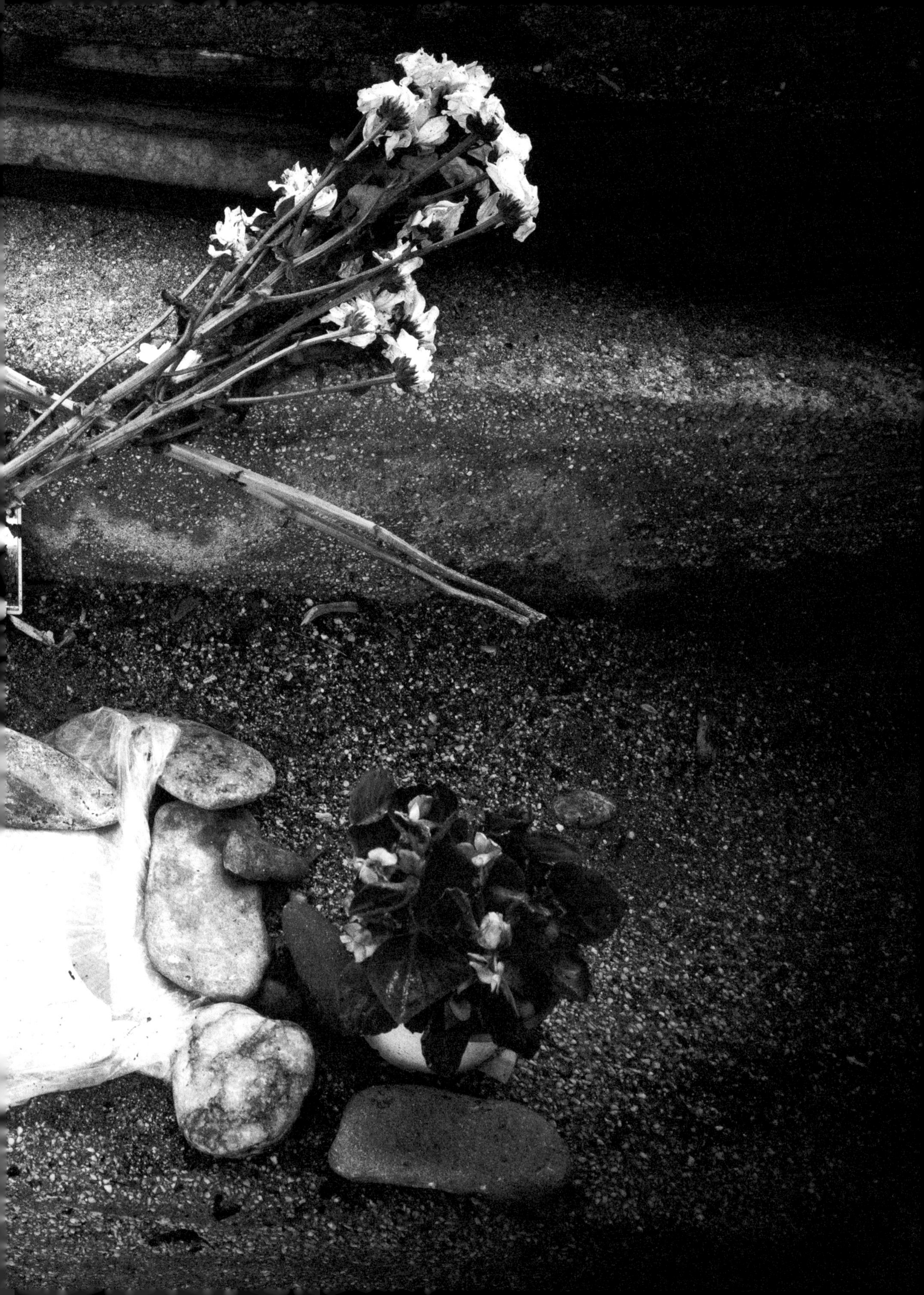

δʹ

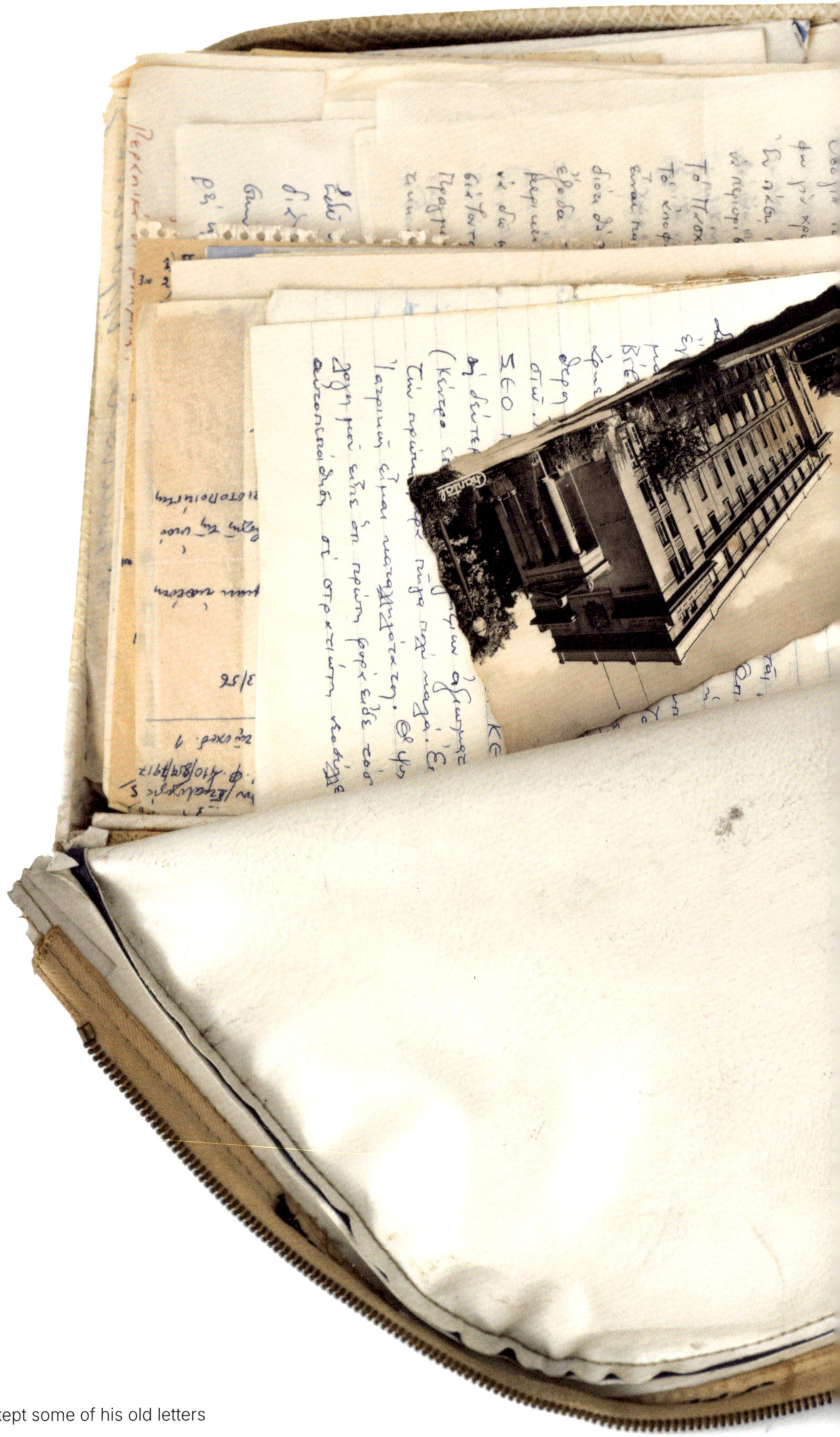

Suitcase where my father kept some of his old letters

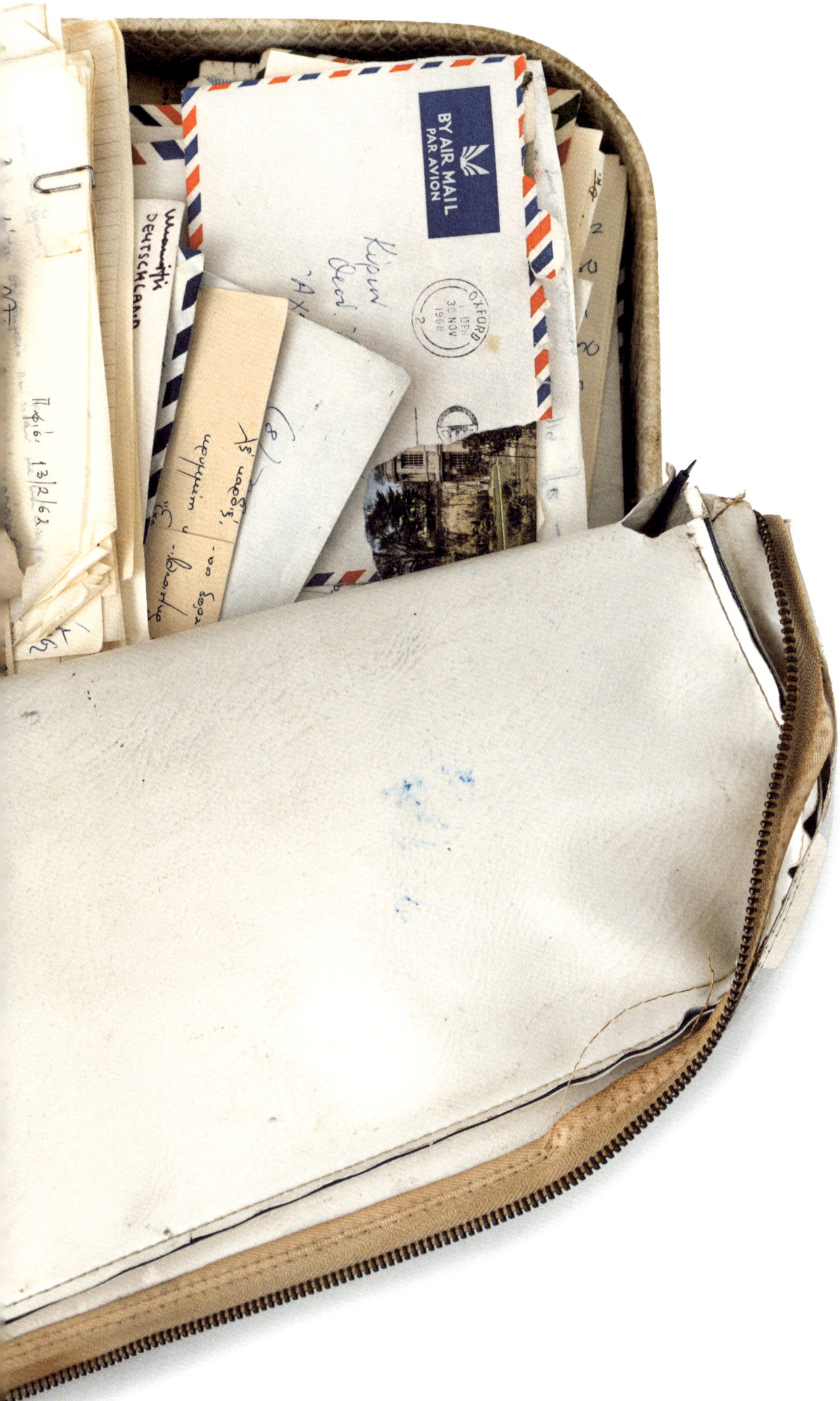
BY AIR MAIL
PAR AVION
OXFORD
30 NOV 1968
DEUTSCHLAND

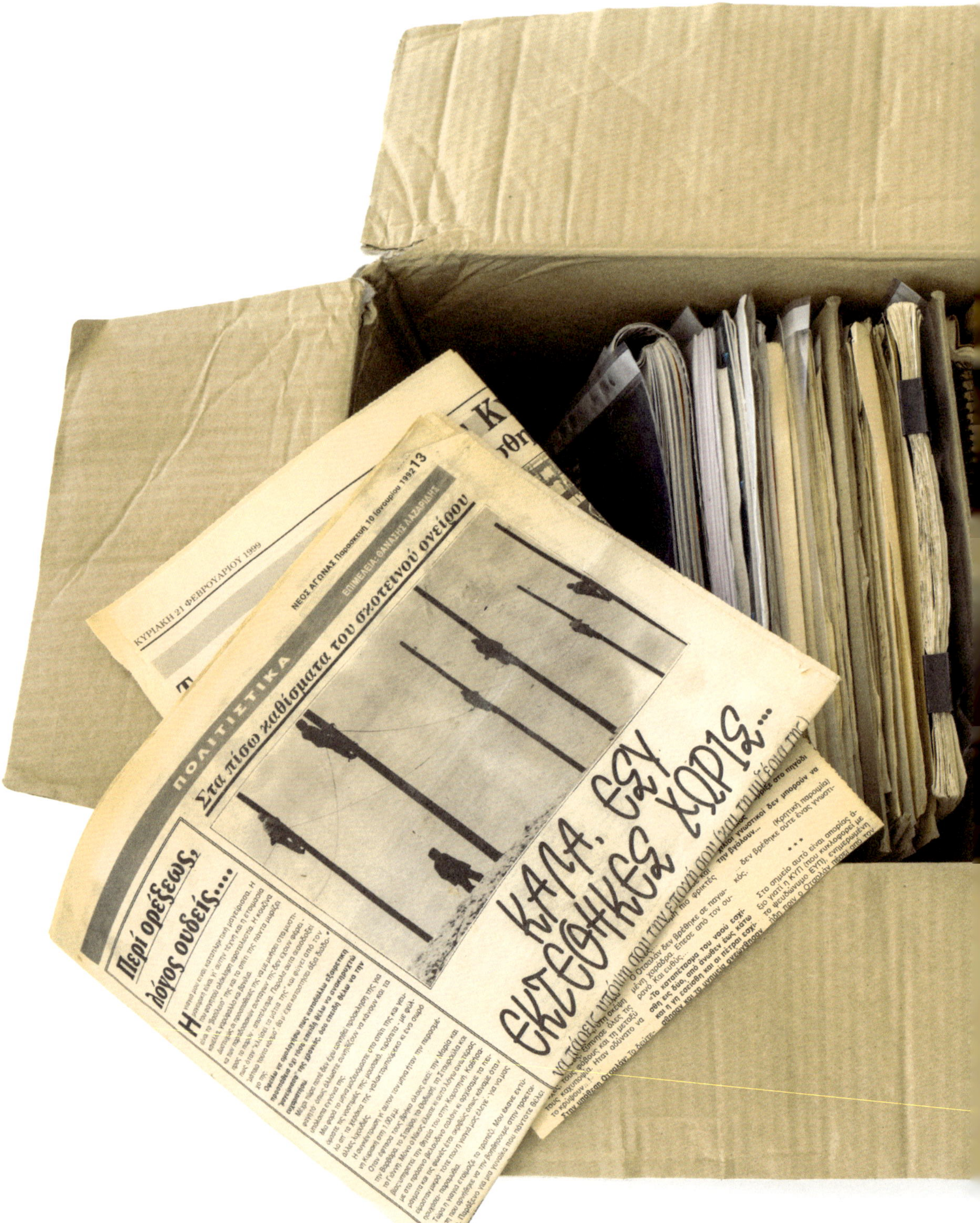

One of the boxes my father kept newspaper cutouts for his work and some of his writing

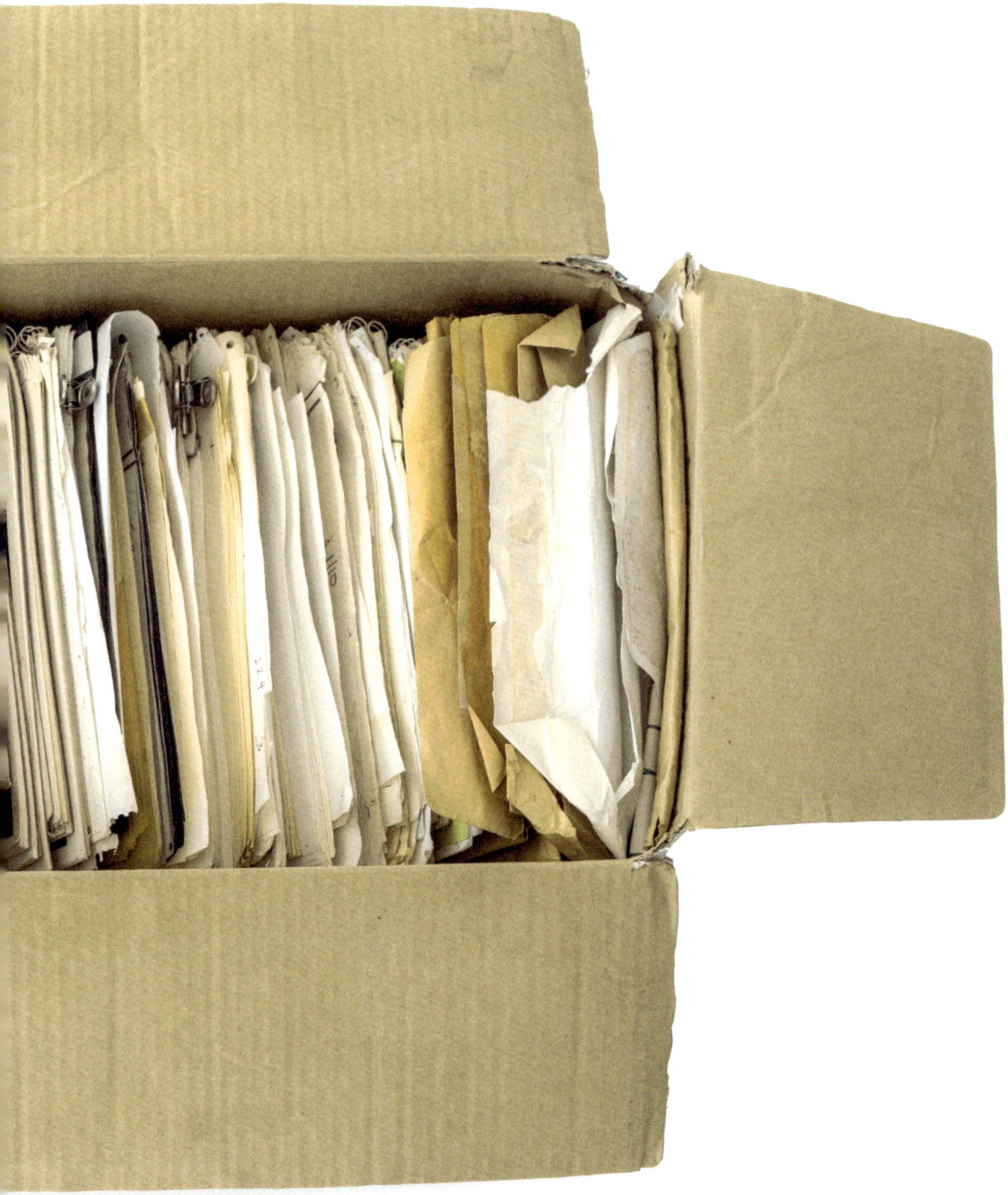

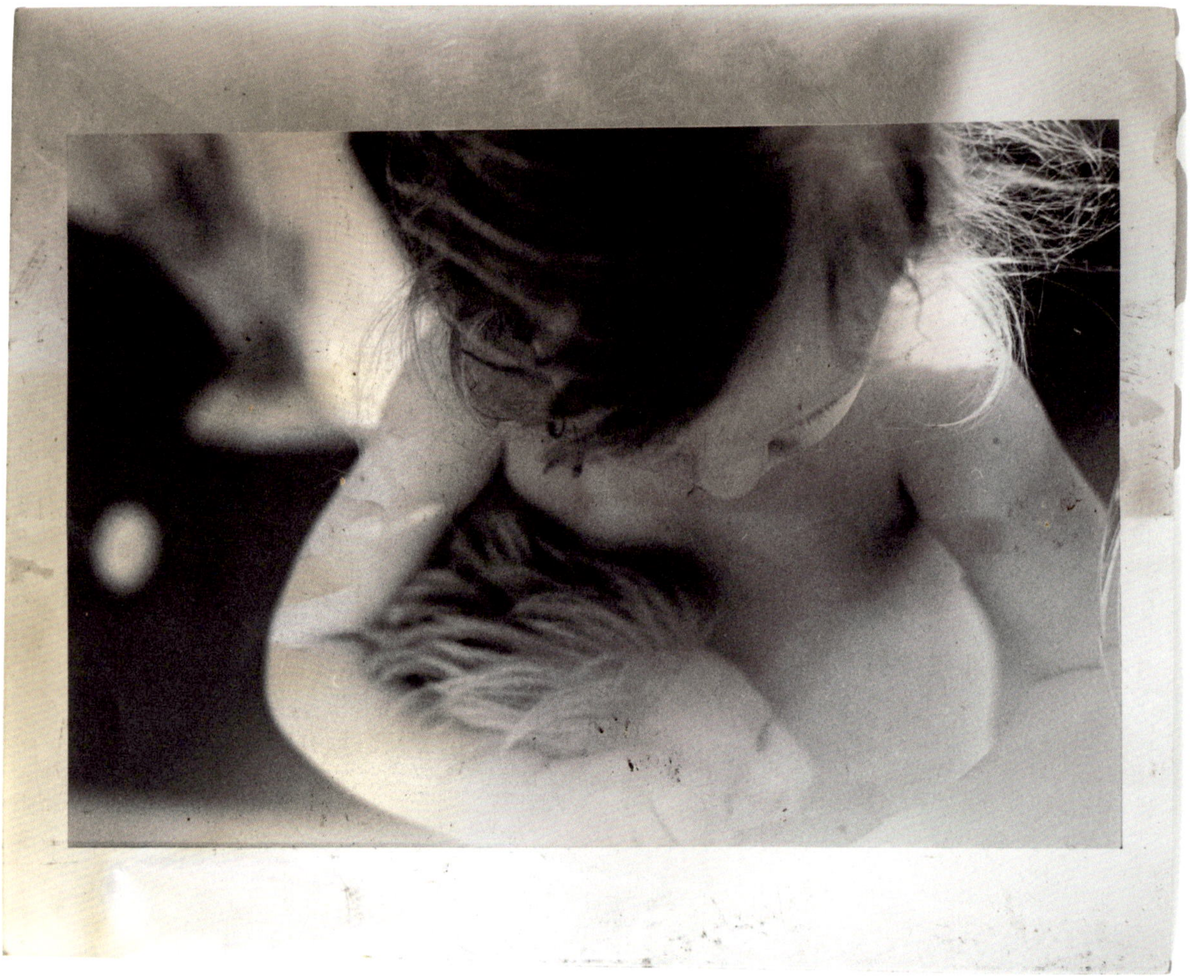

Black and white darkroom prints of Clio found in a room that survived the fire.

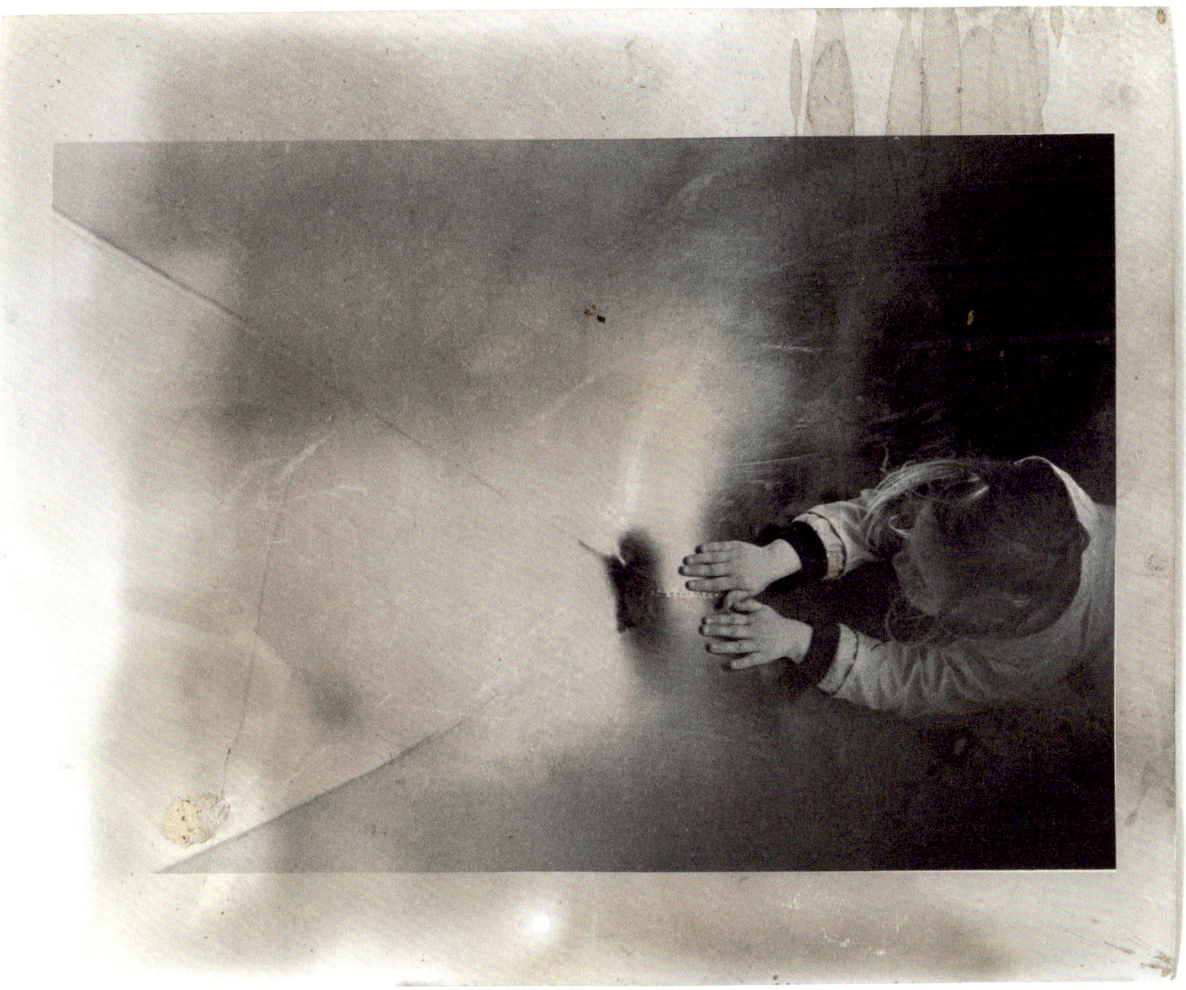

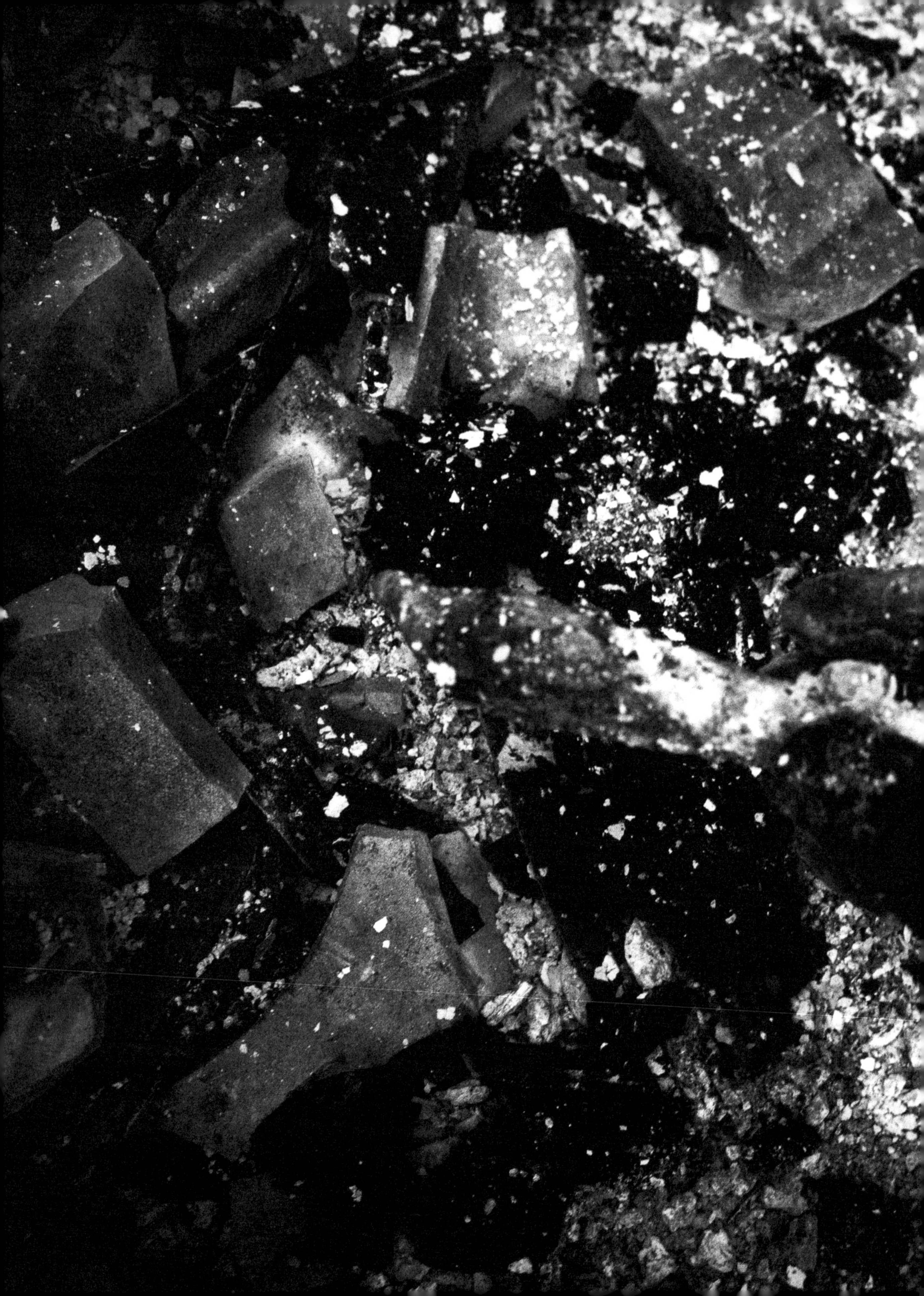

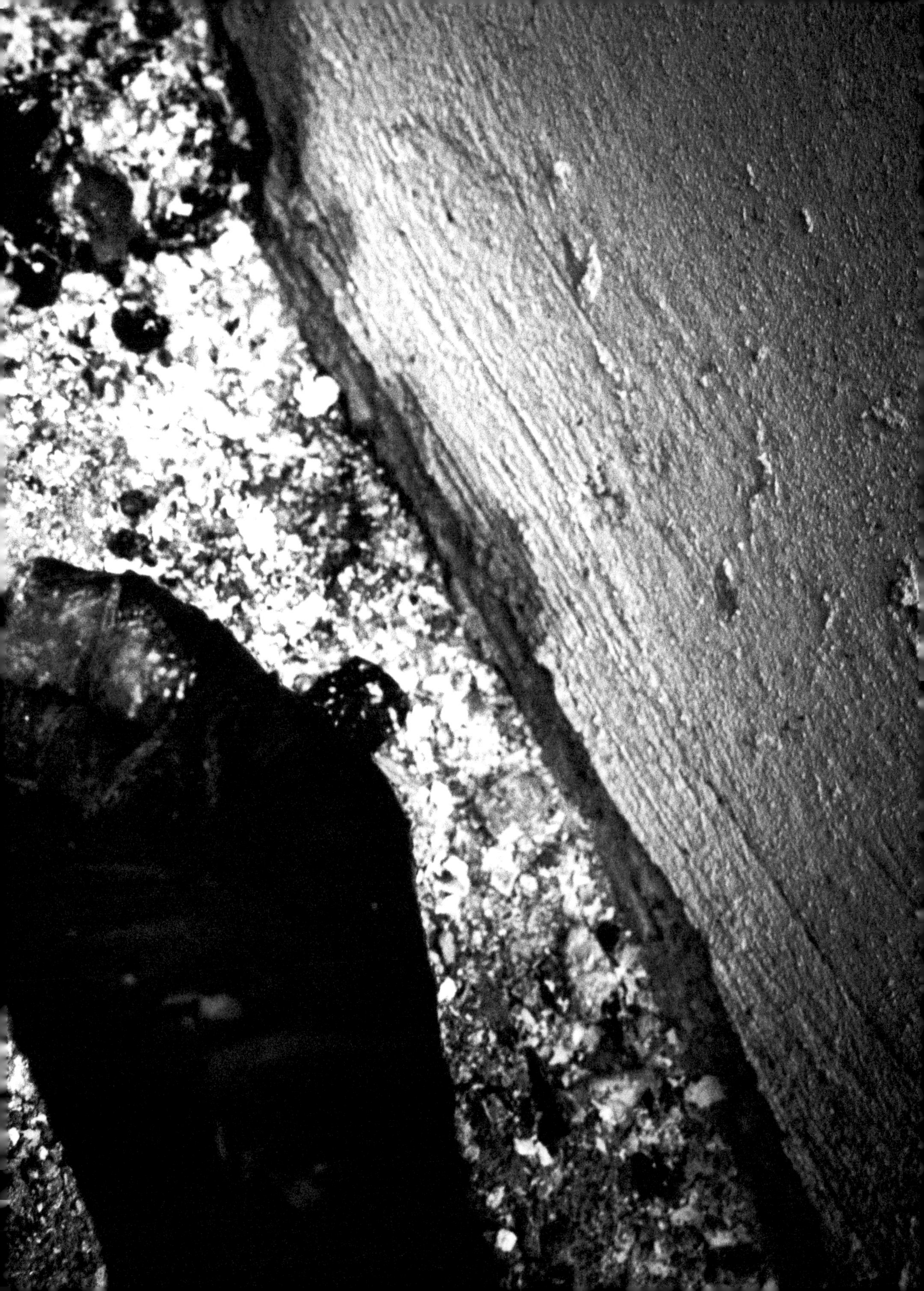

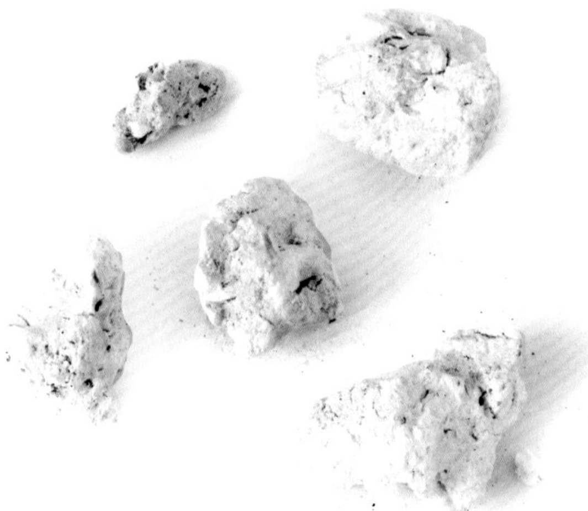

Fabric from clothing

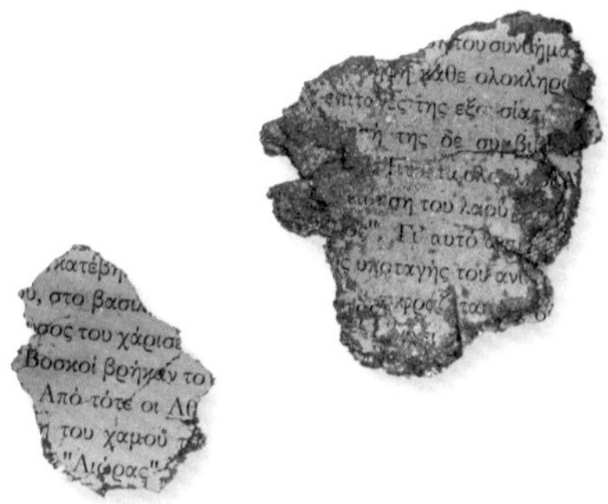

Excerpt from a book

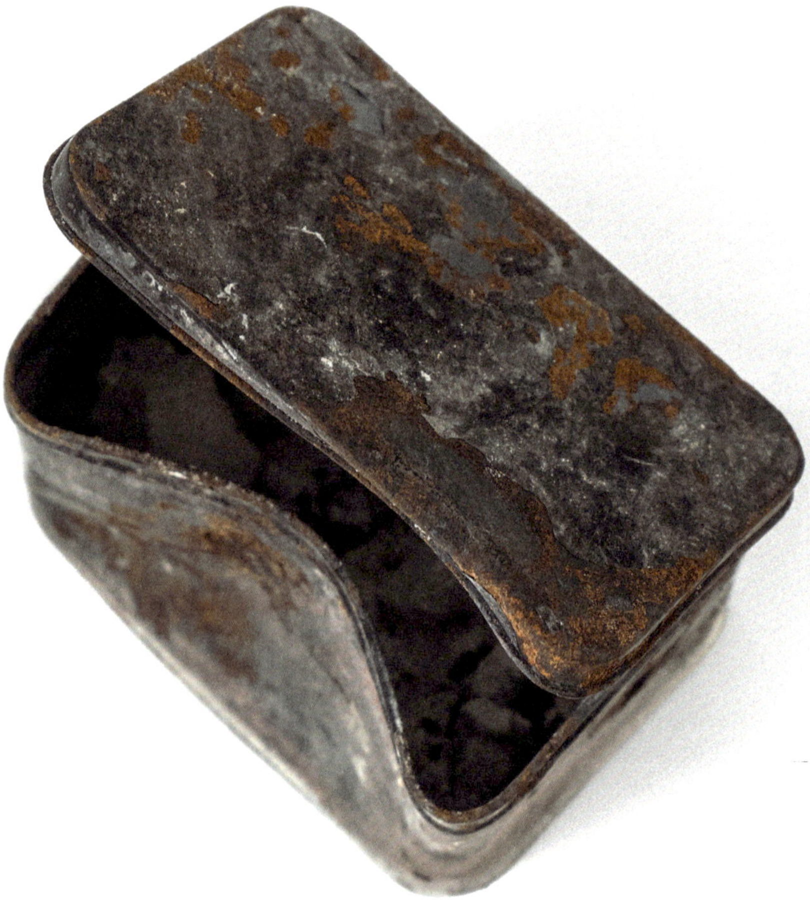

Tin containing notes from school friends

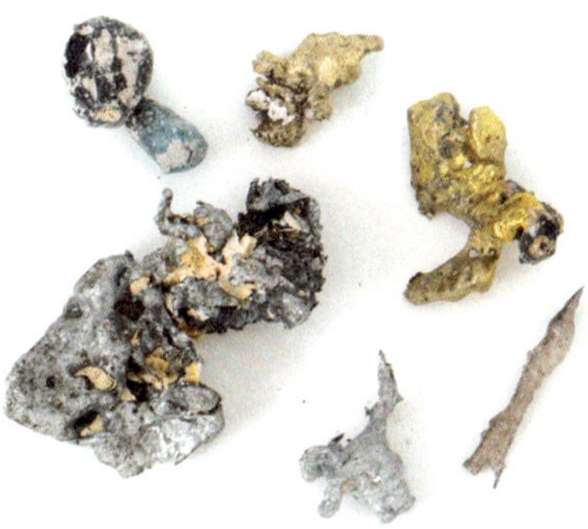

Gold and glass from jewellery

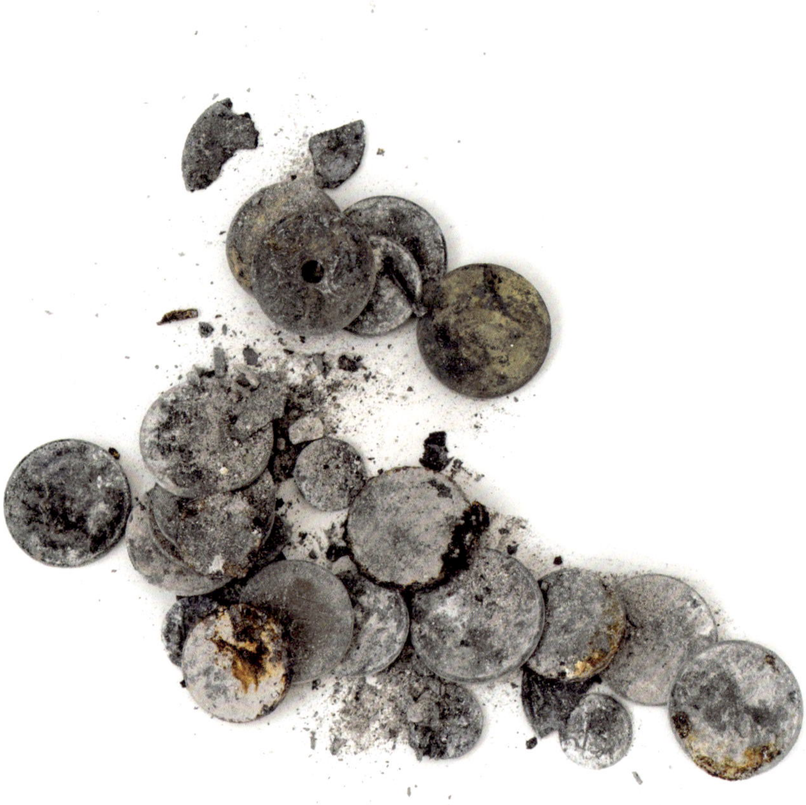

Coins

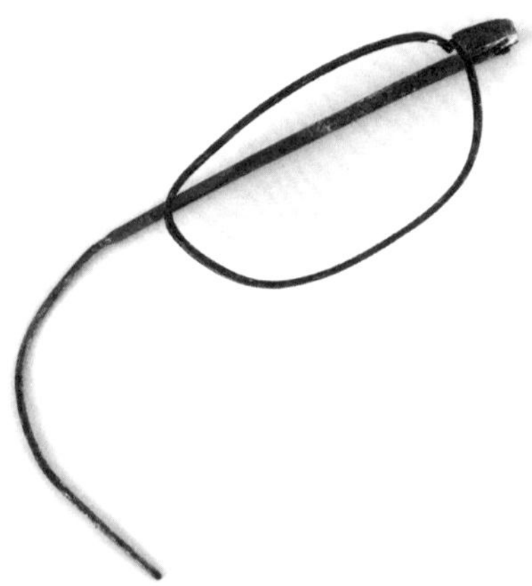

Glasses

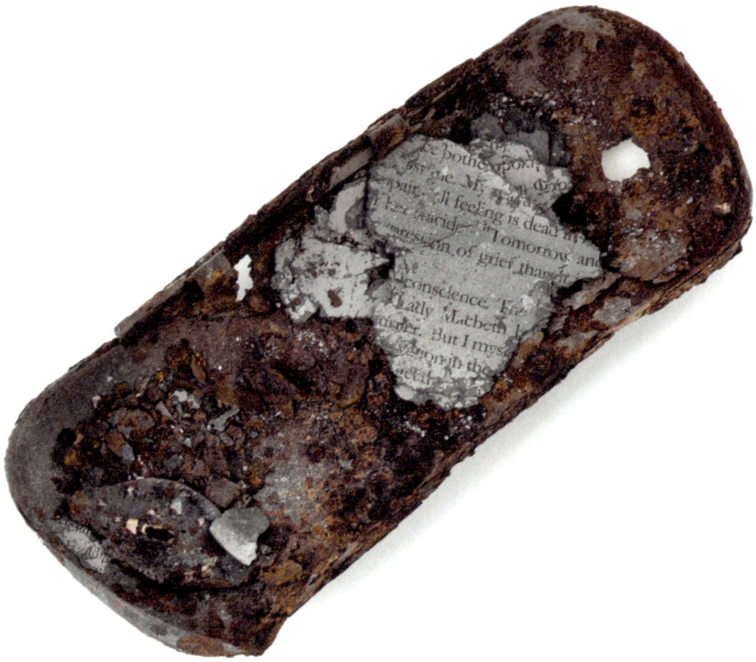

Glasses case and a piece of paper containing
what's left of William Shakespeare's *Hamlet*

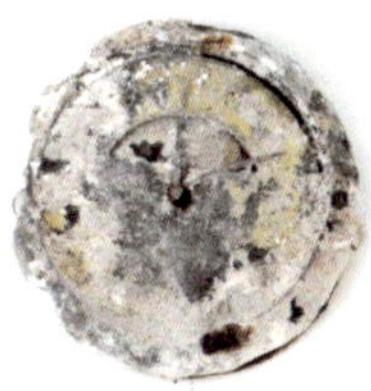

Wristwatch

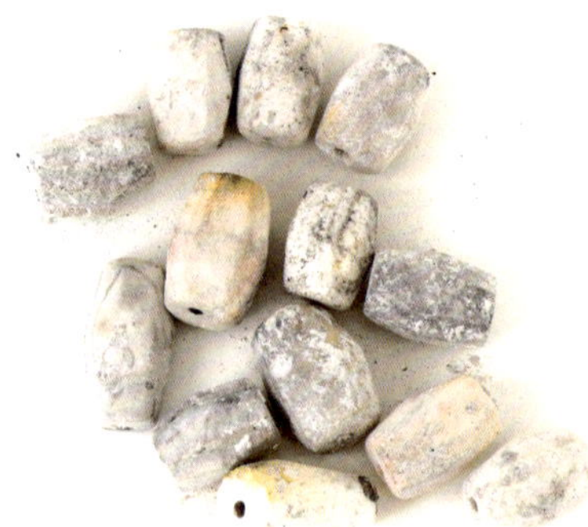

Beads from a necklace I bought from Portobello Road, when I first moved to London in 2000

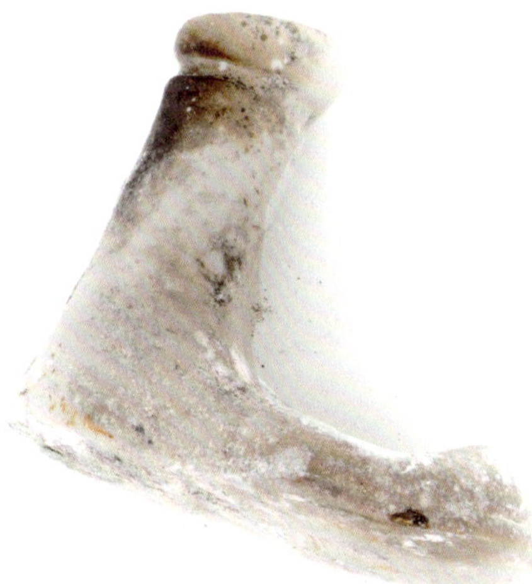

Ceramic boot from a hand puppet my
mother bought me when I was little

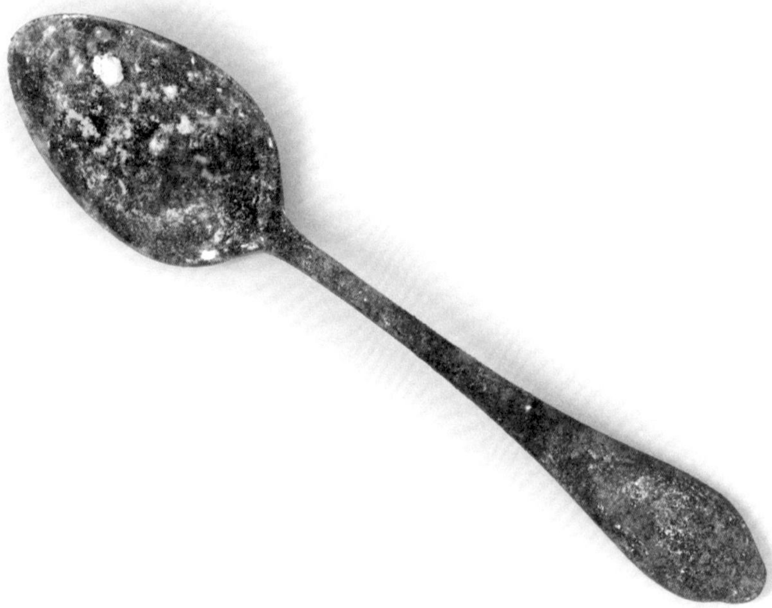

Spoon

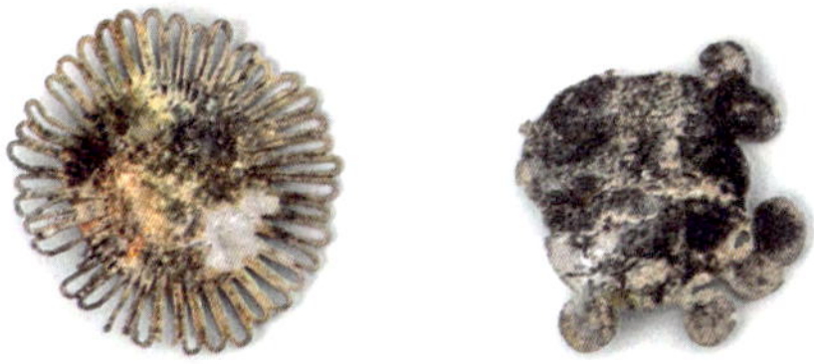

Earrings

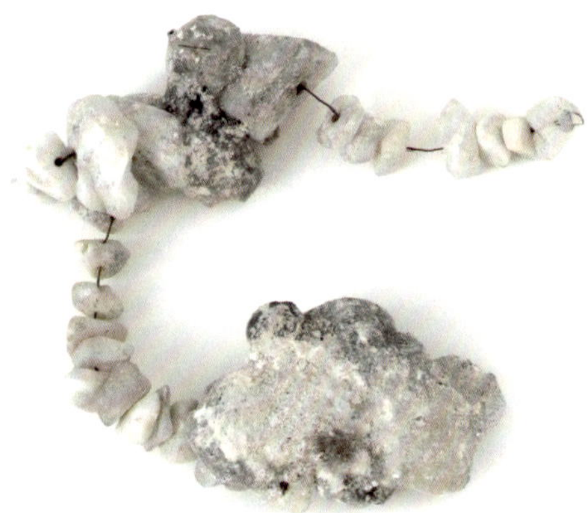

A bracelet of seashells made with my daughter
on the beach the year before the fire

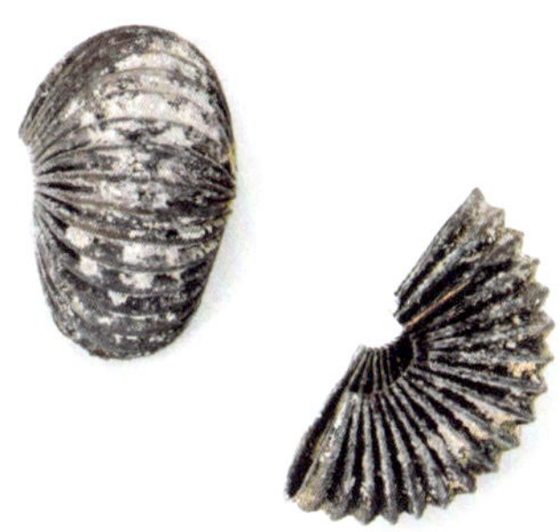

Pair of earrings, bought from vintage shop for a play in 2008

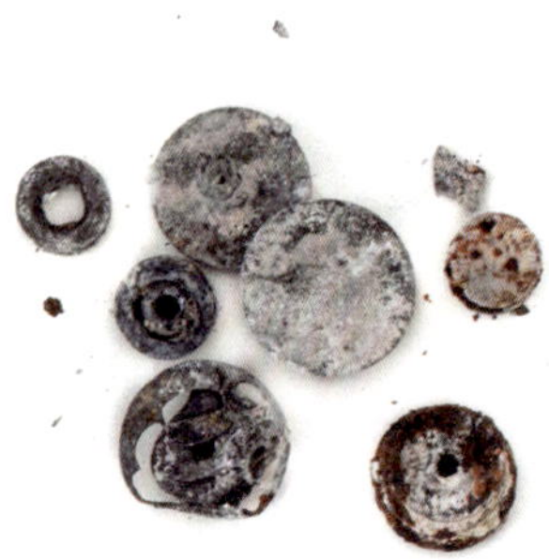

Various buttons

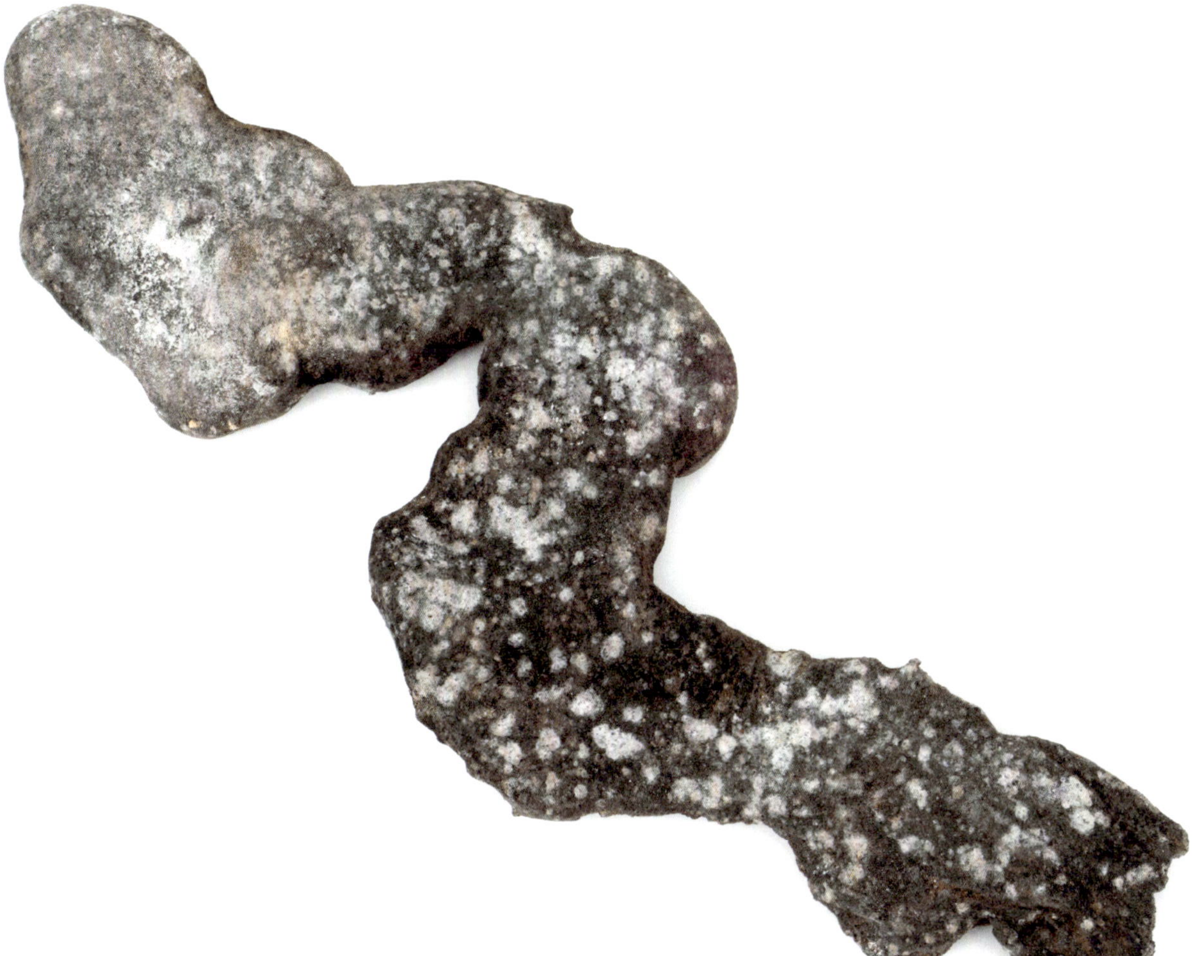

Car alloy

3

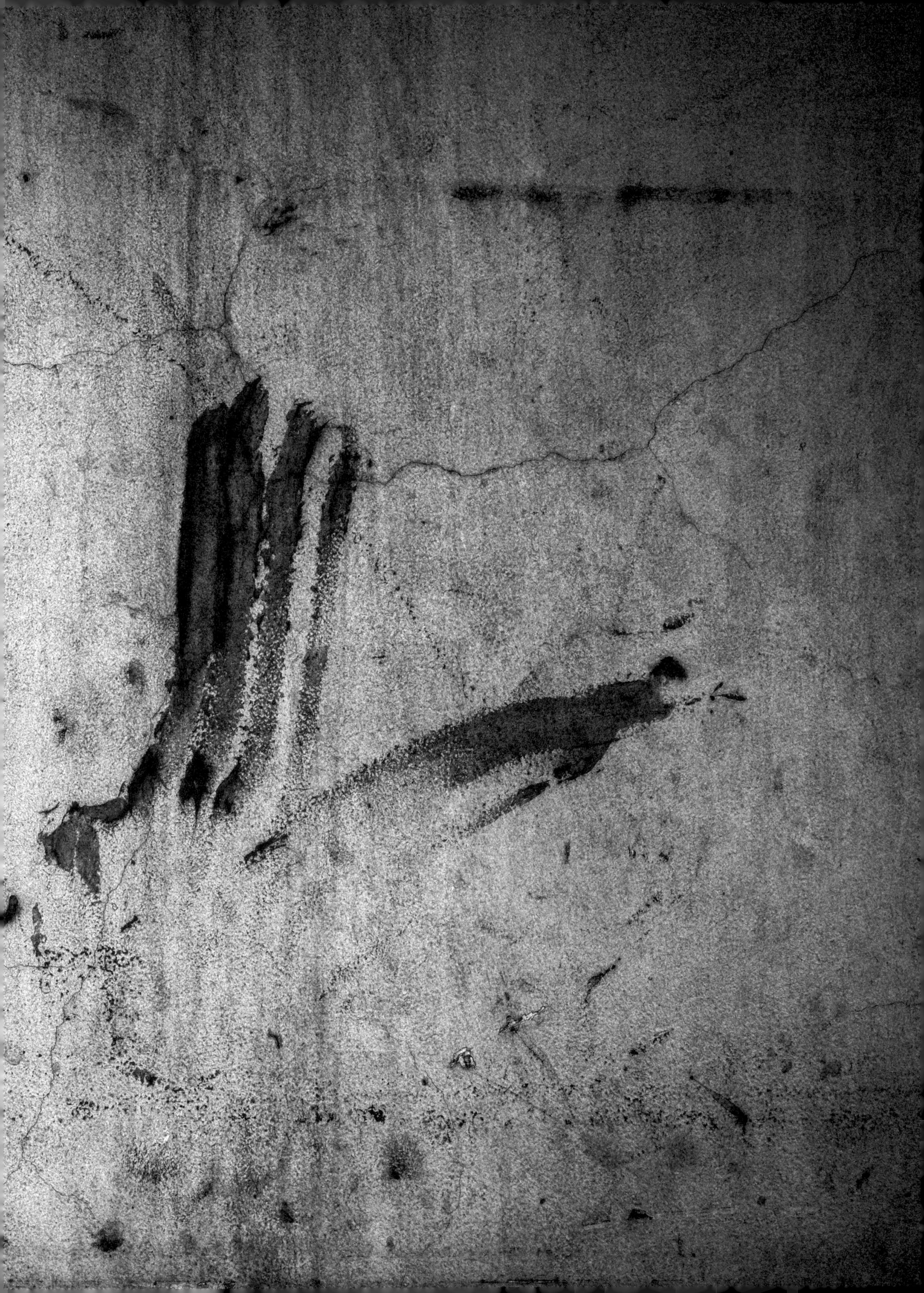

Afterword
Katerina Angelopoulou

This is the way the world ends
This is the way the world ends
This is the way the world ends
Not with a bang but a whimper.
—T.S. Eliot

A large wildfire has a very distinctive sound. No one can tell you unless they have been in one. It is a sound that can haunt you; like the sound of the earth during a big earthquake. What overwhelms you is the force which is no other than the force of nature, in reaction to the wounds we inflict on her.

On the 23rd of July 2018, in the North East of Athens, and 30km to its historical centre, one of the deadliest wildfires ever recorded took place. It was in the area I had spent my summers and some of my winters as a child, the home of my parents, my father's personal office, and ultimately the only place I could identify as 'Home' in my ever-transient existence. I had arrived from New York, where I was living at the time, two days earlier with my three-year-old daughter.

The temperature was 40ºC and wind gusts reached 100 km/h. There was mostly silence around, as it was still resting time in these sleepy seaside villages. The fire rushed from the hilltop, jumping over the pine trees, and to the edge of the sea in a little more than two hours, swallowing everything in its passage. Trees, people, houses, cars.

The Greek Fire Department, the Civil Protection Agency and all other emergency services failed to respond to the emerging situation as well as its immediate aftermath. An evacuation plan, or even a rushed warning was never executed and the necessary forces to battle the fire never dispatched to the area. By 6:30pm the fire had spread to the east, and engulfed the towns of Neos Voutza, Mati and Kokkino Limanaki. Locals and accidental passersby were abandoned by those same structures that were there to protect them. State officials never took any responsibility.

We were at Aimi's house—my mother's best friend, a second mother to me and one of the victims—that day my daughter, my mother and I.

It has been a long and complicated journey since then, to emotional and mental stability, to acceptance of what happened, reconciliation with one's own actions and intimate thoughts, with one's own guilt and with the need to survive, and to survive well, if not for anything else but for the obligation to one's own child.

Since that day, I occasionally photographed what was our home. It took me a long time to be able to go back, and spend time there. It took me even longer to bring my daughter back.

The sounds, images and urgency of that day have never left me. They keep coming back, unexpected visitors, intruders; and take me back to that day, trying to answer the question *why*? Perhaps never to be answered.

This is my journey. If anything, a journey of the mind, a journey through which no moment in time is defined by the present moment. A journey where past and present are permanently intertwined, and through which a new witnessing is possible. The making of an 'object'—a photograph—that can bear witness to one's own experience; to the collective experience; to the collapse of the response mechanisms of the state, and to a more universal experience of climate and other disasters.

The format of the following journey uses principles from auto-ethnography as a method for qualitative research, but also as a means of revealing the process of working. The text takes the form of a partially fragmented and discontinuous narrative, that itself mirrors that of the structure of trauma.

*

The summer (of 2021) started with one of the most destructive wildfires in the history of Greece, yet again. In the early morning, the air in Athens smelled of chemicals as an oil plant had been burning overnight. But that cleared. Then the skies were covered in smog. The haze covered the sun, the light was orange and red. There was that peculiar sense of doom, and the air smelled of fire. It was raining ashes. The wind was howling.

A wildfire in the edge of the Peloponnese and West Attica was burning through a mountain range covered in forestland. It burned through 17.545 acres of land, of which 12.850 acres were forest. No other wildfire in the history of Greece had burned such an extensive piece of land in the months preceding June.[1]

This fire season, as it has been happening across the Mediterranean-type climate regions (MCRs), which are distributed over five continents, started early. Increased numbers of wildfires across the MCR regions have been widely observed, discussed and their images circulated across the world. Scientists have observed temperature increases in response to anthropogenic emissions, and danger seems to increase in regions where precipitation is expected to decrease.

This is particularly true in the Euro-Mediterranean basin, where climate models project both a strong warming—the so-called Mediterranean amplification—and precipitation deficit that collectively exacerbate fire weather conditions. There is already emerging evidence of an increasing frequency of compound heat waves and drought episodes across Mediterranean regions in the observational record and the confluence of the background warming trend with the occurrence of unusually dry years might alter the likelihood of extreme fire danger.[2]

The observed experience in Greece has been that every winter is milder than the last one; and every summer hotter and longer than the one before it. Global warming has been shown to increase the frequency and severity of fire weather conditions in both observed and simulated data, contributing to a lengthening of the fire weather season as well as an increased frequency of days with elevated fire danger.[3]

Population shifts, as well as expanded building has brought inhabitants and homes into the wild land-urban interface (WUI). In combination with climate conditions and (what is usually) anthropogenic ignition, they have resulted in extreme wildfire events (EWEs), defined as very intense fires that often result in very large burned areas with significant impact on human lives and assets.[4]

As wildfires become more frequent, and the amount of land, structures, wildlife and human life that are affected or perish increases, there is another chronic (health) issue that is developing—that of trauma.

¶

I walk towards the house.

There is nothing real about what is going around me.
I am calm—I feel Clio and my mum are safe
for the moment.
I have left them with a baby raft and a chair
by the beach.
I was given some clothes and changed Clio.
She's less lethargic.
Her body temperature must be normal again.

The air has cleared out a little.

There is absolute silence.

Everything is dark, black, burned out cars,
and scattered low fires burning.

An elderly man, with his wife beside him,
has suffered burns. We bring water.

They are calm, silent, patient...

I walk further up.

The house is on fire.

The sight is breath-taking. I turn around and leave.

How could I have just left?

I was told later that I called out for my father.

I have no recollection of it.

Maybe in my head, I lost my father again that day.
I feel I did.

And, I called for him, the same way I called that day when I realised that it was he, who was lying in the tarmac in front of me, hit by a motorcycle.

I remember asking 'Why, dad?' Another why.
Again, a why.

I had this childish superstition that his spirit would have protected the house—along with everything that was left of him—and us.

I guess he did what he could.

He protected us.

But not everyone was lucky.

Little did I know at the time, about what had really happened.

Little did I know that I walked past bodies of people that had been caught up in the fire; little did I know that the smog had concealed sights of indescribable suffering and death; little did I know that the horrendous sound of the fire and the explosions had concealed their screams of fear and agony.

I still don't know; what I saw.[5]

*

> **τραύμα** το [trávma]: κάθε βίαιη διακοπή της συνέχειας του δέρματος ή των ιστών του σώματος ανθρώπου ή ζώου.[6]

Following the translation of the Greek definition of trauma, we read: 'Every violent interruption in the continuation of the skin or the tissue of the body of a human or an animal. What distinguishes the use of the word trauma—at least in the English language—in modern medical or psychiatric literature, is that trauma is understood as a wound inflicted not just upon the body, but upon the mind. It is a breach in the mind's experience of time, self and the world.[7]

A breach, which unlike bodily trauma that most probably heals over time (albeit fire burns often continue hurting for years), does not.

What distinguishes an event that is experienced and then stored into our memory, from an extreme event that is experienced as trauma, is the way that it is experienced at the time of its occurrence. As Caruth specifies in her analysis of trauma narratives and history, 'the pathology of trauma consists solely in the structure of its experience or reception: the event is not assimilated or experienced fully at the time, but only belatedly, in its repeated possession of the one who experiences it.' And she specifies, 'To be traumatised is precisely to be possessed by an image or event.'[8]

I called my mother to make sure she stayed indoors

with windows closed to avoid breathing in any of the smog. She had instead, gotten into the car, and driven as far as possible and close to the sea so as to not smell the fire at all, exactly as she had done almost three years ago in the fire of the 23 July. She had driven back to her house that had burned down. The smell and smog setting her off in a complete alarm as they had reached all the way there. She was experiencing difficulty breathing. She seemed not to understand what and why this was happening. She started screaming...

In more recent years, and with the development of science and medical imaging, scientists came to understand more about the inner workings of memory, and the distinction between assimilated 'narrative' memory and non-experienced, non-assimilated 'traumatic memory.' The two being differentiated by their social and temporal adjustability.

Narrative memory is produced when the mind follows modes and schemas of its past experiences and is able to categorise it. With time these memories can become inaccurate and distorted when 'new ideas and pieces of information are constantly combined with old knowledge to form flexible mental schemas.[9] On the other hand, a traumatic memory, like the one of the fire of the 23 July, is formed when an event or an experience defies categorisation. These memories remain fixed and unchangeable and as Van der Kolk specifically notes in his work with trauma victims, '...the victims relive in the present unmodified traumatic episodes of ten, twenty, or thirty years ago.'[10]

Further study of neuroscience and brain imaging has proven that to be the case, as scans have indeed shown alterations on many parts of the brain; in the dorsolateral prefrontal cortex (DLPFC) and the hippocampus, which are those areas that ascribe meaning and context to experiences. The DLPFC in other words tells us 'how our present experience relates to the past and how it may affect our future.' Van der Kolk further characterises its function by naming it *the timekeeper of a brain.*[11]

Hence, the [past] traumatic event is experienced in its belatedness and again in the present.

¶

DREAM

I am back in my parents' house. It is not burned.

My father is there. He is not dead.

It's summer. Midday.

A fire. It's coming fast. It's already in the back yard.

It is already burning the orange trees.

We are there. We are in a car. In the back yard.

There is fire everywhere.

I am trying to pull my father out of the car...

TRUTH (?)

We are in the car. There is only one-way
to go—towards the sea.
The other way the fire is coming.
I grab my sneakers from the house and
I put them on.
Motorcycles drive past us; people are running.
We are stuck.
There are cars in front us, and behind us.
There is nowhere to go.
My dad's files are in the car; we are not
leaving the car.
The cars are moving.
They are all turning right, towards Rafina?
They are stuck again.
'Go left. We'll take the car through the Bonanos'
house to the "beach."'
We go left. The fire is in front of us. It's burning the
trees in the Bonanos' house. Another car is parked
in front of the wooden gate.
There is nowhere to go.
'Take Clio and run. I'll come.'
I run again. We run. I can't see anything.
I can feel Clio's hand firmly in mine.
I know my way without seeing.
I know where the steps are.
There, 1-2-3, then straight and left.
I pull the rope. I open the gate.
We jump to the rocks.
There are more people there.
We look behind us. Where is my mum?
I can't see again. I can't see her.

I can only see a huge wall of fire.

¶

On the days that followed the fire, the truth of the victims and their families were questioned multiple times. On the days that followed, my truth was questioned multiple times. The media, the systems in control of the public narrative about the fire, as well as the government, concealed facts and re-produced false arguments that blamed the residents and the victims. The government displayed an ever-present image of being in-control, solidarity and care towards the victims whilst at the same time stating that they [the victims] were the makers of their own destiny. They were writing the history of the event on their own terms. The confusion and the cat-and-mouse blame game that was created therefore, assisted in an ever-increasing circulation of this story and false information in the media.

The victims and the survivors had their actions questioned, and as a result they perhaps, themselves started questioning their own actions. An extreme wildfire, which had been left to take its own course, became a life threatening, and therefore, traumatising experience for everyone that was caught in it. A second re-traumatisation took place in the immediate aftermath by a number of media outlets and the government, that were controlling the public narrative, when they refused to witness the 'true' event in its aftermath. The re-victimisation further took the form of raising bureaucratic walls in every attempt of the victims to reclaim their lives.

The fire spread and engulfed the villages, before the media was aware of it. As a result, an outside, objective observer was never able to witness the event whilst it was unfolding. A handful of images of live TV broadcasts were made later that evening, after the fire had almost died down, and during the rescue efforts from the various beaches.

The same people that suffered through this event, some victims, and some survivors, took the only images that now exist. They were the singular witnesses of an event they could not witness.

The survivors in the immediate aftermath had only two ways to tell their story; to make their truth available for the world to hear and to contest the (by then) public opinion. As their voices were not heard, their stories fragmented and sensationalised but not truly witnessed, these images became their only, even if partial, testament to the truth, their trauma and evidence to their re-traumatisation. It was this photographic evidence that demonstrated the very fact of *a crime*. The optical [...] trace became [...] an essential prerequisite for its acknowledgment by society.[12]

I was one of those people, maybe perhaps because I did not know, at the time that I shot those images, what really had happened, what was behind the smog that is not evident in those images. *I wonder whether I would have taken them if I knew?*

What these images show is less than what they conceal. How can an image speak of such horror, but not show it?

*

ich Weiss,
ich Weiss und du weisst, wir wussten,
wir wussten nicht, wir
waren jada und nicht dort,
und zuweilen wenn
nur das Nichts zwischen uns stand, fanden
wir ganz zueinander.
[...][13]

Almost as an internal need to reaffirm the truthfulness of the event, and as if the only way for myself to witness this event and its aftermath, I started photographing. As Baer remarks, while interpreting Meyers Levin's writings, '"the survivors" ordeal did not end with the war but continues in their struggle to become witnesses to their own experiences.'[14]

I photographed the house and my mother; her sadness was unbearable to hold. I photographed my daughter's return to the house, and what once used to be everyday summer rituals and now were laden with an ever-present sense of loss.

Photographing people, even my mother, felt from the first moment, wrong. She did not allow her pain to be visible to the outside world—only as aggression. And therefore, I felt like I would be betraying her. I would reveal something to the world that she had so insistently tried to hide. It would exhibit to the public something that was inherently private, risking the destruction of the singularity of suffering.[15] I felt the same way about others. I would be asking them to reveal their vulnerability. My daughter, being a child and less inhibited, expressed her hatred for being photographed in what remained of the house.

Photographing victims or survivors of atrocities, wars or any kind of violence has been a long-contested territory in photography and photojournalism. These photographs have been long accused of re-traumatising the same people, who in a way the photographers are trying to 'help.' What a 'photographing project'[16] would achieve, would be to expose those same people as less fortunate to the eyes of any audience, and brand them—again—as victims of their own self-imposed (as the public perception has not been changed fundamentally) misfortune.

I needed to find a different language, which perhaps borrowed elements from what we know as 'forensic photography,' but bore *within* it a relationship with time—the relationship that trauma has with time, which is an internal one, a structural one. Additionally, I needed to propose a different reading to the one propagated by the media and government in the aftermath.

Lowe remarks that 'traumatic representation is caught in a paradox of representation' and suggests, 'there is considerable potential in examining images associated with the atrocity that do not depict the actual act of violence or the victim itself, but rather depict the circumstances around which the act occurred.'[17] In this specific event, the landscape, apart from the site of trauma and a victim itself, bore within it the evidence of the circumstance of what I call, the *crime*. It was a crime of negligence and a crime of re-victimisation.

In addition, 'as memory is inherently spatial, so spatiality is inherently temporal, occupying a place in the present but stretching back into the past.'[18]

I spent time walking, retracing my steps, retracing other people's steps according to their testimonies, traced the locations where victims were found, rediscovered the topography and morphology of the area, and stared a lot into the sea. It was *not* a cerebral encounter.

¶

I remember seeing people going into the sea. A group of tourists were boarding a tiny fishing boat. They went further into the sea to be safer. The current took them even further...
I remember after a while I did not see them anymore. They had disappeared in the endless blue that was lying in front of us.
I learned many had drowned—among them a kid and his mother from the boat.

Again, I did not see.

¶

The narrative, taking the lead from the *'optical unconscious'*[19] of the photographs and from the process, develops through disrupted notions of linear time. Locations that were intentionally used by the media to subvert the narrative against the residents of the area, have been re-photographed as themselves sites of trauma, and have been set next to photographs that speak for their innocence.

¶

I read some newfound testimonies...
Covered by the smog.
Covered by the voices and the people that were screaming next to us.
Covered by the fear and the horror, there were people burning alive 20 metres away.
The camera pointed in their direction,
they unseen.
I see now that which I did not see.
I do not sleep...

*

Every turn of this project, every development, was a departure from, but also a return to the time and the site of trauma.

This confirmation of the reciprocal exchange between past and present, not as flashbacks, but existing in the present is what made the use of the photographs of the day of the fire necessary in the narrative. Not as evidence, but as *Historical.*[20]

In using those images, what also is attempted to be resolved is what Baer[21] calls the 'crisis of reference' that photographs can produce. Through the inclusion of those photographs, there is an attempt to reference, as a lived experience, by those present in the photograph or by the photographer. A consideration, which has troubled me in the development of *The Fumes of Mars* and which, in a way, also negates my above-mentioned attempt to a degree, is that none of those photographs—my photographs—represent the event in its full horror.

This attempt to resolve this crisis of reference, which is, however, never fully resolved, as those photographs themselves reproduce it, was mitigated with the use of language. The use of language speaks of certain aspects of the unspeakable. And, together photographs and text potentially provide a reference for the event and for the trauma.

The landscape as the background within which a figure, form, or narrative act emerges, exerts the passive force of setting, scene and sight.[22] The symbolic, real and imaginary space and place of Mati (the chosen point of reference for this event as the fire affected a wider area that included the villages of N. Voutzas, Kokkino Limanaki, and forestland in Pendeli), informed the opinion of every member of the public.

The reading of *The Fumes of Mars,* allows for an expanded interpretation with the nature of the event and the landscape that is not confined within the pages of the book. It questions the persistent established narrative and offers a different reading that highlights the need of an involved audience. An audience that questions its sources of information. It aims to bring to the forefront questions about the nature and structure of Greek society. Why has such an event been allowed to happen, and how willingly people forget. It hints at the ways a society has eroded through the years of economic crisis. And, how this erosion led to an erosion of its structures and the appearance of populism.

The Fumes of Mars uses the 'notion' of trauma itself and its inherent temporal structure, in that it reveals it, as it remains on the psyche of the survivors, and on the landscape.

June 2021

¶

1 Zachariadis, P., Ert News Online, 24/05/2021, accessed at: https://www.ertnews.gr/frontpage/xekina-i-episimi-katagrafi-ton-zimion-apo-ti-fotia-sta-geraneia-oi-ameses-apozimioseis-video
2 Ganteaume, A., et al, Understanding future changes to fires in southern Europe and their impacts on the wild land-urban interface, Journal of Safety Science and Resilience 2, 2021, p20-29
3 Ganteaume, A., et al, Understanding future changes to fires in southern Europe and their impacts on the wild land-urban interface, Journal of Safety Science and Resilience 2, 2021, p20-29
4 Moreira, F. et al, Wildfire management in Mediterranean-type regions: paradigm change needed, Environmental Research Letters, Volume 15, Number 1, 2020.
5 For more in the concept of the "betrayal of sight" and Trauma, see Caruth, C.: Unclaimed experience: Trauma, Narrative and History, chapter 2, and her analysis of the film by Alain Resnais "Hiroshima mon Amour"
6 The Modern Greek Dictionary, Institute for Modern Greek Studies of the Aristotle University of Thessaloniki (Manolis Triantafyllidis Foundation)
7 Caruth, C., Unclaimed Experience: Trauma, Narrative and History, Baltimore, John Hopkins University Press, 2016, p2.
8 Caruth, C., (ed.), Introduction, Trauma: Exploration in Memory, Baltimore & London, The John Hopkins University Press, p4.
9 Caruth, C., (ed.), Van der Kolk, B. & Van der Hart, O., The Intrusive Past: The Flexibility of memory and the engraving of trauma, In Trauma: Exploration in Memory, Baltimore and London , The John Hopkins University Press, p 172.
10 Caruth, C., (ed.), Van der Kolk, B. & Van der Hart, O., The Intrusive Past: The Flexibility of memory and the engraving of trauma, In Trauma: Exploration in Memory, Baltimore and London , The John Hopkins University Press, p 172.
11 Van der Kolk, B., The Body keeps the Score. London, Penguin Random House, 2014, p.69
12 Lowe, P., In Kennedy, L. & Patrick, C. (ed.), The Violence of the Image: Photography and International Conflict, The Forensic Turn: Bearing witness and the "thingness" of the Photograph, New York, Routledge, 2020. My substitution of the word atrocity for crime.
13 Celan, P.. Die Neimanrose, Athens, Agra Publications, 1995.. Original in German. Translation my own: I know. / I know and you know, we knew, / we did not know, we / were there, after all, and not there / and at times when / only the void stood between us we got / all the way to each other.
14 Baer, U., Spectral Evidence; the photography of trauma. Cambridge, MIT Press, 2002, p.94.
15 Dugane, E.. In. Reinhart, M. et al (ed)., Beautiful Suffering: Photography and the Traffic Pain: Photography after the fact, Chicago, Williams College Museum of Art & University of Chicago Press, 2007, p. 97.
16 Rosler, M.. In Wells, L. (ed.), The Photography Reader: In, Around and Afterthoughts, London & New York, Routledge., p.268
17 Lowe, P., In Kennedy, L. & Patrick, C.. (ed.), The Violence of the Image: Photography and International Conflict, The Forensic Turn: Bearing witness and the "thingness" of the Photograph, New York, Routledge, , 2020. My substitution of the word atrocity for crime.
18 Trigg, D., The Memory of Place: a phenomenology of the uncanny, Athens, Ohio University Press, 2012, p.xvii.
19 Benjamin, W., A Short History of Photography, 18/06/2021. Accessed at: https://www.artforum.com/print/197702/walter-benjamin-s-short-history-of-photography-36010
20 Tagg, J., In Wells L. (ed.), The Photography Reader: Evidence, Truth and Order, London & New York, Rutledge, 2008, p.260.
21 Baer, U., Spectral Evidence; the photography of trauma. Cambridge, MIT Press, 2002, p.88
22 Michell, W.J.T. , in Michell, W.J.T. (ed), Landscape and Power, Second edition, The University of Chicago Press, Chicago, & London, 2002, p vvii.

Challenging Images
Michael Herzfeld

Can a bitter tragedy bring about cultural change?

The Mati tragedy illustrates both the repressive power of cultural stereotypes and the redemptive power of those who challenge those images and their constant use as a device for evading blame. In the response to emergency and disaster, Greeks are accustomed to seeing a pattern in which the relevant officials refuse all personal responsibility and deflect it to others. They conventionally represent national-level bureaucrats, not as conscientious public servants, but as self-serving petty tyrants unwilling to risk their job security by taking any kind of initiative.

They know also that formal poses—whether of impartiality or indifference—sometimes conceal interest and engagement.[1] In one version of that scenario, heroic whistleblowers refuse to accept their superiors' willful failure and, at great personal risk, apportion blame where the public believes it should lie. That is what happened in the aftermath of the Mati tragedy.

A public prosecutor, armed with the testimony of one such conscientious official, commented scathingly, 'The indifference shown in combination with the increased responsibility of those who are to blame constitutes a potential threat of escalating the danger in which the citizens found themselves.... Despite knowing all this, the appropriate authorities [of the firefighting service] displayed indifference (*adhiaforisan epidhiktika*) and watched the scene unfold without getting involved.'

Note the double mention of 'indifference'—the second time as a verb (*adhiaforisan*) in a past tense that indicates a specific act, and implicitly a deliberate one, of indifference. While we cannot convincingly read people's minds or claim knowledge of their intentions, the long time over which officials did nothing does suggest intentional inaction. The whistleblower secretly recorded the reactions of his superiors on his cell phone. In one such moment, a fire service official tells him 'to fix some conclusion about the fire, not [to say] that one or another person is at fault...Fix a conclusion using your experience...The Prosecutor won't say anything [critical] to you!' The official then continues, 'Five lines, five words, five whatever...and that'll be it! The lack of aerial surveillance and firefighting equipment emerges as a pattern of deliberate neglect, backed by blackmail: 'Bury it, or we'll rip you to shreds!'[2]

Bureaucrats often face accusations of *efthinofovia* (fear of responsibility). The Greek term carries an abiding sense of deliberate evasion not obvious from the English translation. It evokes an image of the state as a fate-like presence that allocates suffering capriciously. Greeks often talk about fate, but, like people everywhere, they invoke it actively to ward off personal responsibility for failure. 'Fatalism' is a myth, a vicarious quality, attributed to those over whom the speaker has some power as a way of justifying and enforcing that power. For the dominated, conversely, invoking fate—or the power of bureaucratic superiors—is a way of reclaiming agency in the face of failure. Fate works equally well as an excuse for oppression and for being oppressed.

Consider, for example, the way several government officials blamed the Mati residents for illegal construction. Then Minister of Defense Panos Kammenos was unequivocal: 'This is a crime from the past,' he said, 'This coast of Athens, all these properties, the majority are without a licence, and they have occupied the coast without rules.'[3] Such statements preemptively conceal official responsibility. Why had the state neither immediately prosecuted the residents for building illegally nor tried to regularise those illegal constructions as a way of easing the social and economic pressures that had allowed them to spring up in the first place? Neither option would have been palatable to the bureaucrats. Both prosecution and legalisation would have entailed administrative overload and political turmoil.

The pattern is clear. Determinedly retrospective accusations of illegality ('crimes from the past'), cynically deployed to inculpate the residents instead of inactive officials, mark an active idiom of self-exoneration. Such devices are commonplace and instantly understandable by all concerned. Note that I am not saying that the charge is untrue; I am asking *how it is used*— a very different question. Beyond the immediate act of indifference by officials who should have intervened quickly lies a larger structure of morally questionable complaisance—complaisance that may have facilitated the residents' needs at the time, but that then also set them up for the subsequent tragedy.

In defying building regulations, the residents had acted in terms locally understood as normal, even unavoidable. The formal procedures for building legally are time-consuming and expensive. That is convincing as an excuse because the law and the bureaucracy are indeed usually slow to act, while exemplary punishments are rare. Greek cityscapes display clear evidence of ingenious improvisation, a feature of the larger self-image of many Greeks as *katafertzidhes*, fixers. That term has strong intimations of what I have called cultural intimacy—that area of everyday life that officialdom views with disapproval.[4] It is a space, nonetheless, where 'everyone does it.'

'The impact of this situation is both visible and durable.' A Cretan friend once remarked, 'There is nothing more permanent than the provisional.' Those who build illegally can also use the slow pace of legal process to stymie attempts at removing illegal construction already in place. The same excuses, conversely, serve the interests of functionaries deeply reluctant to make unpopular moves against citizens and to deal with the inevitable intransigence, political mobilisation, and simple opprobrium that such interventions tend to trigger.

These observations suggest at least a partial explanation of what happened—and, perhaps more significantly, what did not happen—in Mati. If the authorities thought that the illegal constructions constituted a serious fire risk, was it not because the legal process itself takes too long, and because the residents would have taken full advantage of the range of delaying tactics that the sluggishness of the system made possible? Residents and officials play in the same moral field; they appeal to the same supposed rigidity of 'the law'—in many ways the modern substitute for 'the fates.' The residents proactively protect their dwellings as long as possible; the officials proactively look the other way. What looks like fatalism is actually a socially sanctioned—and unofficially tolerated—exercise of personal agency shared by those with formal power and those who lack it.

The bureaucrats can decry the residents as 'irresponsible,' as they did in the case of Mati, knowing that such language is an effective—because culturally recognisable—device for evading the consequences of their own apparently deliberate inaction. Their inaction was, so to speak, a form of action. The idiom of excuse-making works well in a space of mutual understanding—though often also of intense dislike—between officials and a populace accustomed to lambasting official maneuvering and prevarication. This (unofficial) system works against 'the system.'

Greeks often stereotypically criticise themselves as unwilling to accept responsibility. For this alleged propensity, ironically, they blame vague forces from the evil eye to 'the Turks.' As a *self*-stereotype, the characterisation works widely and well. When Greeks say that they are too inclined to make such excuses, they effectively reinforce those excuses by investing them with the authority of cultural or historical explanation. What do you expect? Why should we be any different in this from the way we are in any other situation?

Hence the observation of the person in charge of dealing with disasters at the Ministry of Infrastructure: 'Mati personifies all the flaws and mischief of modern Greece.' Her office—which did indeed have responsibility for such matters *in general*—supposedly had nothing to do with what happened *on this occasion*. 'Its' (rather than 'her') failure to act stemmed, so the familiar story goes, from an inherited and incurable national deficiency. She blamed the victims by blaming Greekness itself. In so doing, she was also, consciously or otherwise, creating a better chance that no one would ask specifically why the ministry of which her office was part had done nothing to stop the illegal construction in the first place.

There is nothing passive, nothing fatalistic, about such moves. They are ways of saving oneself from punishment or condemnation—and metonymically presenting oneself as a member of the victim-nation as well—rather than trying to rectify the damage. There is safety in widely accepted generalities. Officials who claimed they had not received any orders or equipment were operating in a cultural field where they could expect empathy and support.

At first, perhaps, they persuaded some segments of press and public. Eventually, however, thanks to the whistleblower and one or two insistent journalists, the press began to attack official inaction and ineptness. The sheer familiarity of the excuses, at first a useful cover, subsequently made it easy to mock them and expose their moral and factual hollowness. Leaving aside the (not inconsiderable) role of party politics in apportioning blame at such moments, the consensus that did emerge shows that the populace never simply acquiesced in the officials' all-too-familiar tactics of evasion, nor did it accept as reasonable their failure to act promptly and effectively. By singling out the one or two individuals who took a conscientious stand, the press simultaneously satisfied a redemptive need for individual heroes whose actions contrastively made others' failings all the more reprehensible. Nothing in this story was inevitable.

The heroes stand out precisely because, in disobeying orders, they seem to incarnate another stereotype of Greek character. They have championed individual morality and independence over government by law, challenging the usual stereotype of civil servants as bound to desk and rulebook. The resulting tension reproduces a historical tension in the state itself—a state that claims its origins in the heroic guerrillas known as 'thieves' (*kleftes*, anglicised as 'klefts') whose few successors who continued the same rebellious behavior were reclassified as 'brigands' (*listes*). So, too, the story of Mati has its heroic kleft.

Dimitrios Liotsios, a member of the fire brigade who accepted a prosecutor's mandate to root out the causes of failure, knew, as the journalist Marianna Kakaounaki put it, that he faced a 'difficult mission.'[5] He would have to identify 'the responsibilities of his superiors'—a phrase that in Greek does not merely imply formal duties but hints that those superiors bore substantial blame for the tragedy. As Kakaounaki reports, Liotsios 'accepted [the assignment] without a second thought.' He quickly felt intense pressure to desist—pressure that included direct threats against his person.

Liotsios did not yield. Not only did his courage produce more information; it also demonstrated that not all Greek bureaucrats act in the conventionally expected way. While common tactics of excuse-making were clearly in operation in the aftermath of the Mati fire, the actions of one determined individual as well as Kakaounaki's investigative journalism were especially instrumental in exposing how they worked.

Self-exoneration acts proactively: people build illegally because they see that others get away with doing the same thing, and no one accepts responsibility for the larger failure to enforce the laws.

There is a Greek saying, 'The fish stinks from the head.' The state's own representatives have access to this rhetorical resource because it fits the model of higher authority to whom the lowly functionary turns in hopes of action only to be disappointed. One such individual, on trial for dereliction of his duties, announced, as he apologised to the victims, 'The state has been weighed in the balance and found wanting.'[6] He doubtless hoped that this conventional phrase, made sonorous by its neoclassical Greek syntax, would deflect responsibility away from him and toward that vague but menacingly fate-like entity.

The defendant's statement was not a fatalistic acceptance of government failure; it was a proactive declaration of disappointed helplessness. Claiming that the authorities had failed to put protective measures in place despite the well-known (and widely experienced) dangers of forest fires, he portrayed himself as the helpless victim of generic inefficiency—a description familiar to all the survivors, whether they accepted his pleading or not. Excuses are not necessarily credible; their familiar form makes them socially acceptable. As for the survivors and relatives of those who died, their rhetoric is more directly believable: they want justice rather than revenge. Revenge codes do exist in Greece, but they presuppose at least a nominal moral equivalence (*isotimia*) between the parties—something the survivors would not accord the bureaucracy. When the balance of power is dramatically uneven, as between a government minister and the victims of a tragedy like the Mati fire, there is no such equivalence; government indifference is an abuse of power. The minister who blamed the victims for building illegally operates in an immediately recognisable idiom of self-exoneration ('What can we do? This is Greece!'). The official who claimed that the state had been found wanting was similarly invoking a piece of received wisdom doubtless shared by all concerned.

If these officials hoped that the excuses would work, they were wrong. A different Greece emerged from the actions of those few who had the guts to speak truth to power. Liotsios risked the wrath of superiors who expected him to protect them from blame, and who could perhaps have punished him for his audacious honesty. In so doing, he exposed not only the misconduct of his superiors but also the hollowness of the familiar excuses. He, too, played a familiar role, that of the defiant hero, but in a way that raised real possibilities of change.

It takes moral courage to challenge entrenched structures of blame and responsibility. Other conventions of Greek self-knowledge, however, ensure recognition of that courage. The media reports play up the image of the heroic David who challenges the governmental Goliath; both the hero and his admirers understand that such publicity will make ministerial petty revenge much less likely. If there is an edifying element in this story aside from that of individual courage, it is also the importance of a free press in a democratic society.

The story of Mati is not only a story of cruel indifference and buck-passing. It is also a story of individual moral courage and responsibility, made accessible by aggressive reporting. The press can demand accountability in ways that compromised and sluggish legal procedures cannot. It can rip apart the entangling net of cultural excuses and botched processes that cloud the public's access to information.

Press freedom helps to trigger real cultural change by inviting reflection on the role of cultural values in the way events play out. Culture is always malleable; only diehard nationalists believe that it is static and perpetual. Since the fall of the military junta in 1974, Greece, buffeted by the cruel indifference of international politics, has responded to external pressures and influences by changing in innumerable ways, yet always remaining distinctive in the global context of each given moment. Improvisation, the hallmark of the *katafertzidhes*, is not only a tool of the cunning. Under the name of adaptation, it can also be a way of learning from the past—not of *blaming* the past for the present as in the ministerial evocation of 'crimes from the past,' but of creatively using the fragments of past perspectives to create a morality for the present age.

The inept and the indifferent can no longer hide behind conventional excuses. Revelations of their failure may instead raise visions of a better social order. Creating a more responsive state requires challenging long-established cultural conventions as well as the power structures that these conventions have protected. A few individuals can play crucial roles. This book, with its visual and ethical focus on a tragedy that was far from inevitable, also contributes to such processes. Its images challenge past indifference and demand a present will to reform.

1 For an extended discussion of Greek (and more generally European) bureaucracy in these terms, see Michael Herzfeld, *The Social Production of Indifference: Exploring the Symbolic Roots of Western Bureaucracy* (Oxford: Berg, 1992).

2 https://www.protothema.gr/greece/article/1027983/tha-mazeutoume-kai-tha-se-skisoume-sok-apo-tous-anatrihiastikous-dialogous-tis-tragodias-sto-mati/

3 https://www.bbc.com/news/world-europe-44966121

4 See Michael Herzfeld, *Cultural Intimacy: Social Poetics and the Real Life of States, Societies, and Institutions* (New York: Routledge, 2016, 3rd edition).

5 Marianna Kakaounaki, Kathimerini Newspaper online, published 19/07/2020, Available at: https://www.kathimerini.gr/investigations/1088260/mati-dyo-chronia-meta-to-egklima-kai-i-sygkalypsi/

6 https://www.kathimerini.gr/society/dikastiko/563404630/terzoydis-gia-mati-i-politeia-ezygisthi-kai-vrethike-ellipis/

Mati: Catastrophe, Chaos, Loss and Healing
Yiannis Gabriel[1]

9 January 2025. As I begin writing this, the Los Angeles area is engulfed by a series of raging wildfires. The number grows relentlessly: four, five, six, then seven separate fires, destroying homes and businesses, prompting thousands to flee their homes. International news agencies report: 'It's utter chaos in Hollywood—streets are gridlocked and people carry whatever belongings they can under a blanket of thick smoke.'

A fire chief observes 'the fire burning in the Hollywood Hills is rapidly expanding and progress on containing the fires is zero'. The death toll climbs hourly. From the comfort of my home in a provincial English city, I witness these events on screen. Closer to home, floods have forced thousands to leave their homes, a stark reminder of nature's fury.

Disasters and narrative

A disaster, whether natural or human-made, is a catalyst for immense anxiety, fear and awe. It paralyzes our ability to make sense of events, leaving us gasping for answers in a storm of chaos.

When disaster strikes, uncertainty reigns. What caused it? How much damage will it inflict? Can it be stopped? Like a city under siege, fear escalates at every stage. Victims and responders struggle to cope, often lacking resources, while witnesses shift between shock, anger, and blame-seeking.

Yet, disasters have become routine. In Greece alone, the tragedy at Mati was followed by devastating fires in Evia, Rhodes, and Evros (the largest on record in Europe) not to mention the annual blazes threatening Athens and the already ravaged Attican landscape. Globally, wildfires, floods, earthquakes, and hurricanes devastate communities, while man-made catastrophes like crashes, spills, terrorist attacks, and wars add to the toll. Are these anomalies, or, as sociologist Charles Perrow suggests, 'normal accidents,' inevitable crashes in complex systems when multiple failures interact and reinforce each other, in spite of all efforts to avoid them?

Disasters cause profound damage, both personal and collective. They often seem arbitrary and defy explanation, yet we instinctively seek meaning. Aristotle noted that humans interpret even random events, like the accidental collapse of a statue, as significant (Poet. 1452a). We construct narratives, accident, negligence, crime, omen, divine will, conspiracy, war, to make sense of chaos.

While narratives cannot erase suffering, they help us process trauma, share grief, and find meaning. They may facilitate healing and prevention or, conversely, keep wounds open, trapping us in cycles of repetition. Disasters are woven into stories with clear characters and causes—some political, seeking justice; others cautionary, urging action; some heroic, honouring resilience; others technical, highlighting systemic failures. Competing narratives shape how we remember, respond, and move forward, or whether we forget altogether.

The fire at Mati: A real and a symbolic watershed

Katerina Angelopoulou's book demands that we do not to forget the fire at Mati[2], a fire that claimed 104 lives and left many more physically maimed and emotionally traumatised. In words and pictures, her book recounts the agony of people trapped in their homes, in their cars, in the angry sea, as flames and deadly smoke closed in. It makes us relive the despair of our fellow humans calling the emergency services and being told to wait for help that never arrived. It depicts the chaos and confusion that prevailed as the fire went about its destructive course and the extent of the devastation it left behind. In our times, when the power of the visual image surpasses that of other testimonies, the book offers as compelling a narrative of the disaster as possible, from the perspective of those who lived through it and narrowly escaped with their lives, losing neighbours and friends as well as treasured properties and memories.

The Mati fire laid bare systemic weaknesses: lack of preparedness and equipment, poor coordination among different agencies, calamitous judgements by officials in conditions of crisis, escalating panic as the fire crossed borders and boundaries that were regarded as secure. It also revealed acts of bravery by many individuals who went beyond the call of duty in helping their fellow human beings in their moment of need, and the good sense of others who evacuated their families or those in their care (like the children in the St Andreas summer camp) against the prevailing official and often conflicting advices. It highlighted the anguish of many of those called to address the disaster at all levels, telephone operators, firefighters, local officials and others, as they found themselves abandoning the Mati residents to their fate.

Because of the unprecedented number of fatalities in a resort where thousands of Athenian families spend their summer vacations, the fire at Mati can rightly claim a certain uniqueness as a great symbolic watershed in the collective imagination. Coming in the aftermath of the financial crisis that saw Greece's GDP decline by 24% between 2007 and 2014, its unemployment rate rise from 10% to over 25% and an estimated 36% of Greeks living below the poverty line, the fire and the authorities' calamitous response seemed to epitomize everything that was wrong with Greek society—the dire state of public services, the corruption of politicians, the anarchic building development and, more generally, the feeling that nobody is in control in a society in a state of chaos where everything is rotten and collapsing.

Against this backdrop, the fire unleashed across Greek society waves of indignation, rage and retribution triggered by helplessness, victimization and despair. A variety of narratives surfaced to account for the disaster, to identify those responsible and seek to punish them. Some of those narratives were realistic, drawing attention to actual failings and deficiencies. Others were fantastic or politically motivated, including attempts to blame the victims themselves for not clearing their plots of combustible material or violating different building and other regulations. Angelopoulou's book demonstrates the falsehood of many such claims. It also highlights the feelings of betrayal experienced by those who suffered from the fire and continue to feel that justice has not been done, the fire being an 'open wound,' a collective trauma that refuses to heal.

In the aftermath of the fire, a wildfire of narratives, conspiracy theories, speculation and rumour flourished. Accusations of cover-up and political weaponisation of the calamity proliferated. These have made it more difficult to derive realistic lessons on how to prevent or mitigate future catastrophes. Fires have since continued to afflict Attica and the rest of Greece—7,000 hectares destroyed in the area of Tatoi-Varymbombi in 2021, over 8,000 hectares were destroyed in Varnavas in 2024, among them. Still, unlike Mati, there were no major human losses from any of these fires, and there is some evidence that timely evacuations and calls for international firefighting support in equipment and manpower reduced their impact. Maybe some lessons are being learnt, even when the cauldron of emotions triggered by the Mati fire continues to burn.

Are there any ways to find solace and catharsis in the aftermath of calamity? Healing begins with acknowledging the magnitude of the loss that we have endured, recognising the tragedy that has afflicted our fellow human beings, some more profoundly than others. It is crucial to understand what happened as objectively as possible, free from the immediate anger and outrage sparked by the disaster. This requires thorough and impartial inquiries. If such investigations determine that individuals or organisations bear direct or indirect responsibility, they must be held accountable—not in a spirit of vengeance, but through the due process of law.

Justice alone, however, is not enough to heal the wounds. True healing demands the difficult and painful process of psychological mourning. As Freud argued in *Mourning and Melancholia*, without mourning, loss remains an open wound, preventing us from moving forward. Mourning, however, cannot take place without acknowledging the suffering of those at the heart of the tragedy and recognising both errors and responsibilities. At the same time, arbitrary blame, scapegoating, and the politicisation of disaster—where opposing factions exploit calamity for political gain—only hinder healing. Worse, they obstruct the opportunity to learn from the tragedy and take meaningful steps to prevent its recurrence.

Fires and other major disasters underscore the point that these are collective threats that threaten all of us. They are threats against which we cannot individually immunise or protect ourselves. They can afflict the rich as well as the poor, the young and the old, the healthy and the ailing. Against the prevailing neo-liberal discourses of individualism where each man and woman must fend for themselves, natural and man-made calamities highlight our mutual interdependence and the need for civic institutions that can be respected and trusted. In consumerist societies which atomize and divide us, which view the state as a parasite sucking up resources through taxation, these calamities stress that we need the state, we need civic institutions, and we need solidarity across a society to prevent them and mitigate them. And yet, these are precisely what neo-liberalism and its populist political narratives are currently undermining and corrupting.

Disasters in Religious and Mythological Narratives

Great calamities have shaped human history, with the worst still described as biblical. Myths and religious stories, such as *Noah's Flood* and similar tales from Mesopotamia, Egypt, and Greece, portray disasters as divine punishment meant to cleanse humanity's sins. These narratives remain relevant as we confront the limits of our planet's resources and the consequences of environmental destruction.

Fire myths are also widespread. In Plato's *Timaeus*, an Egyptian priest tells Solon that the Greek story of Phaethon, the reckless youth who scorched the earth while driving the Sun's chariot, reflects real cosmic cycles of destruction, the movements of the sun and other celestial bodies (Tim. 21e). The Greek version, for its part, attributes the disaster to human arrogance against persistent warnings. This resonates today as climate change worsens fires, floods, and extreme weather events that threaten all life on the planet.

Disasters in scientific and political narratives

Religious and mythological stories have helped people find meaning in suffering. In modern times, science has pitched her own narratives against religious and mythological ones seeking to explain earthquakes, pandemics, climate change, and other disasters with knowledge that was unavailable to earlier civilisations.

However, science does not operate in a vacuum. Political and economic interests shape how scientific narratives are received, often contesting and modifying them. During the COVID-19 pandemic, medical recommendations were debated and adjusted before they became public policy. Similarly, scientists have warned for decades about global warming, yet each new catastrophe is treated as an unforeseen shock rather than the predictable outcome of ignored risks.

Environmental advocates, scientists as well as campaigners, often describe disasters as warnings, yet their lessons rarely last. Victims are left to rebuild while society quickly moves on. Each new catastrophe is dismissed as an isolated incident instead of part of a larger pattern. Many somehow convince themselves that disaster happens to others and will not affect them or their loved ones. This common type of denial is one that psychoanalyst Irving Yalom has investigated with great insight.

From Catastrophe to Renewal: Rethinking our Relationship with Disaster

Disasters, whether caused by nature or human actions, have always been part of human experience. The ancient Egyptians recognised that even chaos had a role in the universe. Seth, the god of storms and destruction, was feared yet acknowledged as a necessary force within the cosmic order.

Like them, we must accept that disasters will happen, but we can take steps to reduce their impact. Science offers solutions, such as those outlined in the Goldammer Report[3] on wildfire prevention in Greece, but these measures require commitment from governments, public engagement, and global cooperation.

Yet, science alone is not enough. Tackling environmental challenges requires coordinated political action, greater public awareness, cross-sector collaboration, and sustained pressure to address climate change. We must work to reduce its impact on human lives and develop creative solutions to adapt to future environmental threats. This effort goes beyond policies and strategies—it also requires new ways of thinking, new narratives and counter-narratives that challenge dominant narratives shaped by powerful interests.

These counter-narratives should emphasise the positive impact of collective action, no matter how small, while exposing the falsehood of those who resist change. They must foster solidarity across communities, bridging divisions and disagreements. Most importantly, they must break the illusion that no alternatives exist and inspire a shared vision for a better future. We need new stories—stories that reimagine how we coexist, both with nature and with one another.[4]

The fire at Mati has left behind a trail of devastation. It remains an open wound for many of our fellow citizens. Only by remembering and honouring their terrible plight during the calamity and the injuries it left in its aftermath, can we move beyond mourning, beyond scapegoating and recrimination to true healing and learning. This work is a step in that direction.

1 I would like to thank Phoebe Oikonomopoulou and Katerina Angelopoulou for their insightful and heartfelt comments on an earlier draft of this chapter. They helped enrich and enhance this chapter.

2 While the fire is generally referred to as the 'Mati fire', it afflicted several other adjoining municipalities. It is known as the Mati fire because it is the area that sustained the heaviest losses and also because it is a scenic resort in which many Athenians spend their summer vacations. The author spent several summers in the late 1950s and 1960s at Mati, in a now burnt house.

3 This deliberations of the Global Fire Monitoring Center headed by Johann Georg Goldammer are available online at https://gfmc.online/allgemein/press-release.html and the full report can be found at https://gfmc.online/wp-content/uploads/FLFM-Greece-Committee-Report-07-February-2019.pdf. They should be read extensively and implemented at speed.

4 There are many environmental success stories that do not receive the recognition they deserve, yet they highlight the transformative power of collective action and environmental advocacy. These stories challenge the established narratives that seek to maintain the status quo. In the case of managing catastrophic fires, several counter-narratives challenge the dominant approaches often driven by political or economic interests. These counter-narratives emphasize sustainable fire management, community resilience, and the role of nature in mitigating fire risk. Notable examples include the Australian Aboriginal Firestick Alliance, the Wildfire Resilience movement in California, community-based firefighting and resilience initiatives in countries like Chile, Portugal, and Spain, as well as fire-resistant landscaping and urban design efforts across various countries.

References page 8–97

1 Information from the State Investigator & Forensic Expert Reports.
2 24hours, Alpha Cyprus Channel Aired on 27/07/2019 (times code: 35¨59 & 1:08:00)
3 Athens Voice online, BBC interview viewed on 25/05/2021, Available at: https://www.athensvoice.gr/politics/463486_proklitiki-synenteyxi-kammenoy-sto-bbc-gia-tin-tragodia-se-mati-kai-neo-voytza
4 www.eleftherostypos.gr viewed on 24/05/2021 Available at: https://eleftherostypos.gr/ellada/270386-martyria-sok-esvina-kati-pou-kaigotan-apo-tis-pyrkagies-itan-mia-nekri-gynaika/
5 Urban Fires Case Study: Mati, Eastern Attica, Greece 23.07.2021 and its Management. Personal Rescue and Emergency evacuation through Survivor's testimonies. Master Thesis. National and Kapodistrian University of Athens, School of Sciences, Department of Geology and Geo environment. George G. Spanos, Special Publications no2020303, Athens March 2021 (pages 40, 41)
6 Private conversation
7 Proto Thema online, published 20/07/2020, 07:55 Available at: https://www.protothema.gr/greece/article/1028097/tragodia-sto-mati-to-ihitiko-dokoumedo-tis-sugalupsis/
8 Star Channel online, viewed on 24/05/2021 Available at: https://www.star.gr/eidiseis/ellada/430952/gnwsth-syggrafeas-ezhse-th-fwtia-sto-mati
9 Kathimerini online, published 19/07/2020 https://www.youtube.com/watch?v=npNqo3b28lc
10 CNN online, Published 17/05/21 Available at https://www.cnn.gr/ellada/story/266186/katapeltis-to-porisma-gia-to-mati-egklimatiki-ameleia-sta-oria-toy-doloy
11 The Guardian online, viewed on 24/05/2021 Avaiilable at: https://www.theguardian.com/world/2019/jul/20/greek-fires-one-year-on-103-dead-survivors-and-rescuers-look-back
12 iefimerida online, Published on 23/01/2023, Available at: https://www.iefimerida.gr/ellada/mati-katathesi-thalassa-pateras-polemos
13 Giannis Papadopoulos, Kathimerini Newspaper online, Published 25/07/18, Available at: https://www.kathimerini.gr/society/976636/i-fotia-kynigoyse-anthropoys-pos-egklovistikan-oi-pyropliktoi-sto-mati/
14 Giannis Papadopoulos, Kathimerini Newspaper Online, Published 19/07/2019. https://www.kathimerini.gr/pages/mati/1035078/i-zoi-meta-ti-foniki-pyrkagia/
15 Thodoris Antonopoulos, Lifo Magazine online, 19/01/2019, Available at: https://www.lifo.gr/now/greece/oi-treis-psarades-poy-elaban-tin-elliniki-ithageneia-miloyn-gia-ti-zoi-toys-kai-tis
16 Marianna Kakaounaki, Kathimerini Newspaper online, published 19/07/2020, Available at: https://www.kathimerini.gr/investigations/1088260/mati-dyo-chronia-meta-to-egklima-kai-i-sygkalypsi/
17 To Ethnos online, published 23/07/2020, Available at: https://www.ethnos.gr/ellada/116791_mati-mama-thelo-na-koimitho-gia-panta-i-sygklonistiki-martyria-gia-ti-fotia
18 Giannis Papadopoulos, Kathimerini Newspaper Online, Published 19/07/2020, Available at: https://www.kathimerini.gr/investigations/1088262/mati-dyo-chronia-meta-katalavame-oti-den-eimaste-asfaleis-poythena
19 iEfimerida online, Published on 07/03/23, Available at: https://www.iefimerida.gr/ellada/mati-diki-liotsios-entoli-ekkenosis-apeiloyn
20 Kathimerini Newspaper Online, published. 25/07/2018 Available at: https://www.kathimerini.gr/society/976761/dramatiki-katastasi-sto-nekrotomeio-athinon-iatrodikastis-protognori-katastasi-maziki-katastrofi/
21 SkaiNews online, Published on: 31/07/2024. Available at: https://www.skai.gr/news/greece/diki-gia-to-mati-eythynes-kai-adiaforia-episimainoun-ta-thymata-stis-katatheseis
22 Banking News Online, Published on 02/12/2022, Available at: https://www.bankingnews.gr/index.php?id=654538
23 Mina Moustaka, Ta Nea Newspaper online, published 25/01/2021, Available at: https://www.tanea.gr/2021/01/25/greece/mati-nea-apokalyptika-stoixeia-kaine-tin-pyrosvestiki/
24 iEfimerida Online, Published 01/04/2010 Available at: https://www.iefimerida.gr/news/488831/zitoysa-na-me-afisoyn-na-pethano-i-teleytaia-varia-egkaymatias-apo-mati-sygklonizei-me
25 NewsIt Online, excerpt from Press Conference of the 26/07/2018, 3/8/2018 Available at: https://www.newsit.gr/videos/toskas-ethesa-tin-paraitisi-mou-ston-prothypourgo-kai-den-tin-ekane-dekti/2587686/
26 SkaiNews online, Published on: 30/07/2024. Available at: https://www.skai.gr/news/greece/diki-gia-to-mati-sygklonistikes-katatheseis-apo-martyres
27 Propaganda online, www.propaganda.gr, viewed on 24/05/2021 Available: https://popaganda.gr/stories/pyrkagia-rafina-mati/

I would like to express my deepest gratitude to everyone who helped make this book possible.

To each and every one of those who supported the funding of this book.

To Stu for his belief and persistence in this work.

To all the GOST team, Gemma, Rossella, Justine, Katie, Claudia, Ana for all their work and expertise.

To all the people who have spent the last seven years researching, writing, and fighting for justice after this disaster.

To Thomas, whose unquestionable support and expertise has been what kept me going.

To Clio, whose bravery and force for life have been a guiding light.

To Vangelis and Evangelos for trusting me.

To Margarita, George, Jo & Dimitris, Sharon, and Theodora for being there every step of the way.

To Alexis for his valuable feedback.

To Elizabeth who supported me through the birthing of this book.

To Peter who was a fervent supporter of this work since the first day.

To Paul, who nudged me to engage with the subject and who persisted. I wish I could share this book with him now.

To Brad for connecting some vital dots, and for being an advocate.

To Yiannis and Michael for trusting me with their texts.

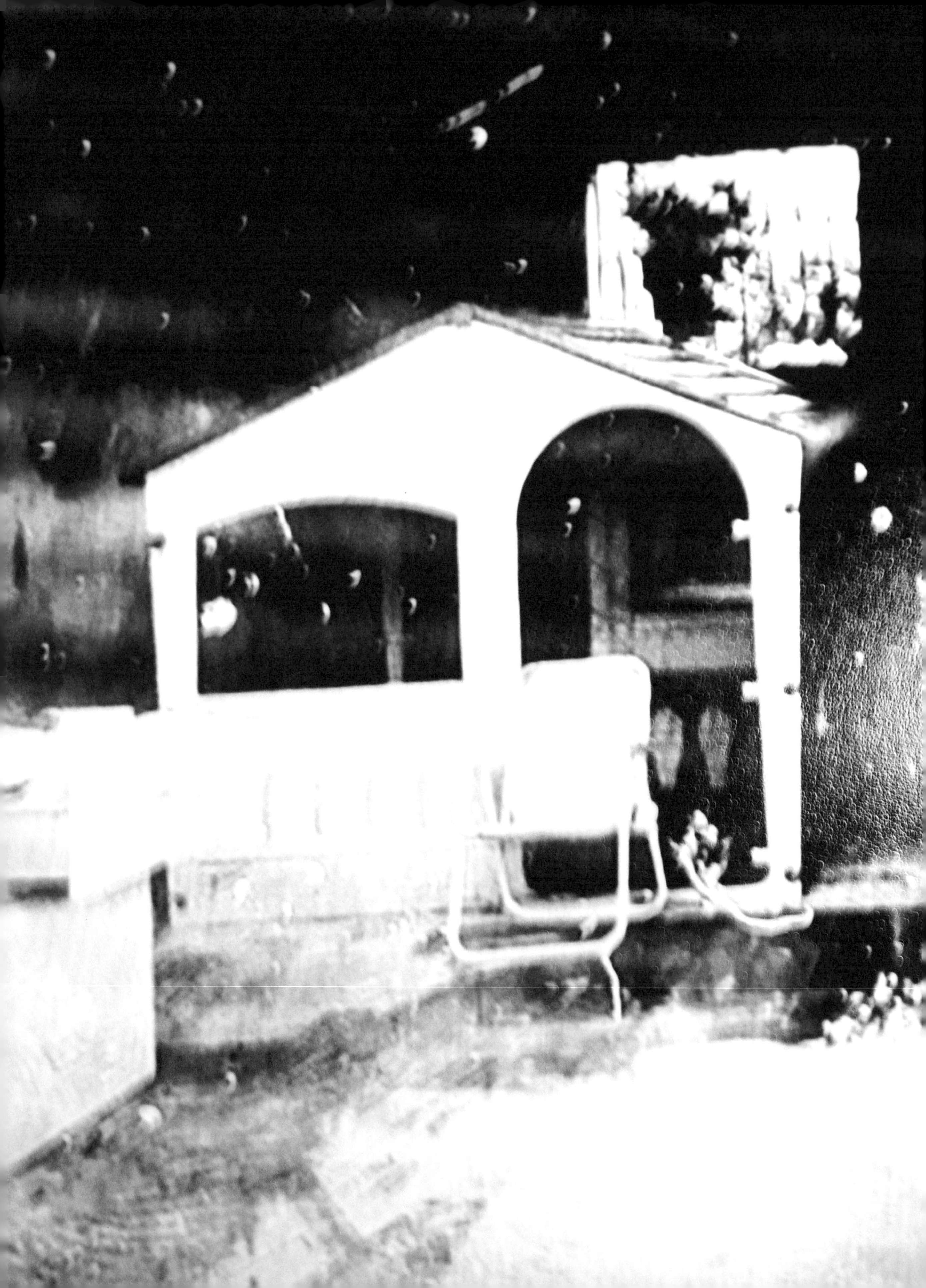

The Fumes of Mars
info@gostbooks.com
gostbooks.com

Edited and designed by GOST:
Rossella Castello, Katie Clifford, Gemma Gerhard, Justine Hucker, Allon Kaye, Eleanor Macnair, Claudia Paladini, Ana Rocha

Printed in Italy by EBS

British Library cataloguing-in-publication data.
A catalogue record of this book is available from the British Library.

ISBN 978-1-915423-88-7